NATURAL HEADLAND SAND BYPASSING: TOWARDS IDENTIFYING AND MODELLING THE MECHANISMS AND PROCESSES

Cover image :
An illustration of sediment dispersion at headland.

NATURAL HEADLAND SAND BYPASSING:
Towards identifying and modelling the mechanisms and processes

DISSERTATION

Submitted in fulfilment of the requirement of
the Board for Doctorates of Delft University of Technology

and

of the Academic Board of the UNESCO-IHE
Institute for Water Education

for
the Degree of DOCTOR
to be defended in public on

Tuesday, June 9, 2015 at 15:00 hours
in Delft, the Netherlands

by

Mohd Shahrizal bin AB RAZAK

Master of Engineering in Hydraulics and Hydrology
UNIVERSITI TEKNOLOGI MALAYSIA, Skudai, Malaysia

born in Pekan Pahang, Malaysia

CRC Press
Taylor & Francis Group
Boca Raton London New York

CRC Press is an imprint of the
Taylor & Francis Group, an **informa** business
A BALKEMA BOOK

This dissertation has been approved by the
promotor : Prof. dr.ir. J.A. Roelvink

Composition of the doctoral committee:

Chairman	Rector Magnificus TU Delft
Vice Chairman	Rector UNESCO-IHE
Prof. dr.ir. J.A. Roelvink	UNESCO-IHE/Delft University of Technology, promotor

Independent members:

Prof. dr. A.H.F. Klein	Federal University of Santa Catarina, Florianopolis, Brazil
Prof. dr. R. Ranasinghe	Australian National University, Canberra, Australia / UNESCO-IHE
Prof. dr. ir. A.J.H.M. Reniers	Delft University of Technology
Prof. dr. A.K.A Wahab	Universiti Teknologi Malaysia, Skudai, Malaysia
Dr. A. Dastgheib	UNESCO-IHE
Prof. dr.ir. A.E. Mynett	Delft University of Technology, reserve member

First issued in hardback 2018

CRC Press/Balkema is an imprint of the Taylor & Francis Group, an informa business

©2015, Mohd Shahrizal bin Ab Razak

Published by:
CRC Press/Balkema
PO Box 11320, 2301 EH Leiden, the Netherlands
Email: Pub.NL@taylorandfrancis.co.uk
www.crcpress.com – www.taylorandfrancis.com

ISBN 13: 978-1-138-37335-8 (hbk)
ISBN 13: 978-1-138-02864-7 (pbk)

To my caring mother Mrs. Norzainab binti Ismail &
firmly father Mr. Ab Razak bin Abd Aziz

ABSTRACT

Natural headland sand bypassing: Towards identifying and modelling the mechanisms and processes contributes to the understanding of the mechanisms and processes of sand bypassing in artificial and non-artificial coastal environments through a numerical modelling study. Sand bypassing processes in general are a relevant but poorly understood topic. This study attempts to link the theory and physics of sand bypassing processes which is significantly important in definition of coastal sedimentary budget and consequently in coastal management. The main questions are how can we model sand bypassing processes and if the modelled sand bypassing processes represent the actual sand bypassing processes.

In this study, it is shown that a process-based numerical model can be used to simulate the processes of sand bypassing around groyne and headland structures for both short (several days) and long term (a year) simulations. Result comparisons were made among analytical models, empirical models and field data. In general, the process-based model can produce reasonable results.

The results of a hypothetical process-based model for a single groyne case indicated that the shoreline evolution patterns and sand bypassing rates are in agreement with the results of analytical models. The shoreline patterns behind the groyne structure are well predicted by the process-based model and represented the reality. Results of the shoreline pattern behind the groyne obtained from the model with the inclusion of wave-groups is better than the results of the model without the wave-groups. The main components that influence the longshore sand transport, *i.e* wave heights, wave angles, and sediment grain sizes, contribute to the sand bypassing processes and different sand transport rates at a groyne structure.

The results of a hypothetical model for a case of two separated groynes showed that the morphodynamic characteristics of embayed beaches can be predicted by the process-based model. Results obtained from an empirical model are consistent with the results of the process based numerical model. Surf zone rip currents, particularly headland rips, are responsible for the sand transporting mechanism outside the surf zone area.

Findings from the field survey analyses at a small nourished embayed beach which is located on the east coast of Malaysia revealed that wave climate seasonality leads to alternating behaviour patterns of the nearshore beach profiles. The results of the cross-shore profile variations along the beach showed the classical onshore-offshore movement patterns of sand transport. Beach rotation, as a result of seasonal wave directional changes promoted sand leakage at one end of the embayment. Additionally, results from the sand volume analyses at the southern area outside the nourishment zone positively showed the contribution of sand into this region and thus verified the potential growth of sand southwards around the south headland.

For the southern Gold Coast case study, a process-based model of a permanent sand bypassing system was introduced by utilising the concept of a sand bank discharge operation. Prior to morphological model investigation, models were successfully calibrated and validated and model results were compared against the available field data measurements. The results obtained from the morphological models revealed that additional supply of sand from the sand discharge operation contributed to the development of a sandspit. The sand bypassing process began when the sand started to move around a headland as sandwaves. The sandwaves manifested themselves, moving slowly in front of the beach, and creating an elongated sandspit. The sandspit grew bigger and bypassed another headland before finally attaching itself to the neighbouring beach. Results from the qualitative assessment showed that the morphological beach behaviour captured by the process based model represented the actual morphological beach behaviour as in reality. The modelled sand bypassing process in this study is identical to the conceptual model of headland sand bypassing. A final conclusion is that the combination of seasonal wave climates, in particular wave directions, and the sand bank discharge operations determined the succession of the permanent sand bypassing system.

In summary, through numerical modelling this study has added to the understanding of coastal processes and the role of geological controls in governing sand bypassing processes and embayed beach morphodynamics. The morphological model developed in this study is useful to increase understanding of the natural sand distribution patterns due to combination of engineering efforts and natural coastal processes.

SAMENVATTING

Natuurlijk zandtransport langsheen kapen: Naar de identificatie en het modelleren van mechanismen en processen levert een bijdrage tot het begrip van de mechanismen en de processen die leiden tot het natuurlijke transport van zand langsheen landtongen in milieus die al dan niet beïnvloed zijn door menselijke ingrepen. Hierbij wordt gebruik gemaakt van een numeriek modelinstrumentarium. Het transport van zand voorbij kapen is een heel relevant, maar weinig begrepen vraagstuk. De voorliggende studie tracht een link te leggen tussen bestaande theorieën en de achterliggende fysica, met tot doel het verbeteren van de bepaling van sedimentbudgetten. Dit moet leiden tot een meer verantwoord kustzonebeheer. Dit werk gaat in op hoe zandtransport langs landtongen kan worden gemodelleerd, en op hoe goed die modelresultaten overeenstemmen met veldgegevens.

In deze studie wordt aangetoond dat numerieke procesmodellen aangewend kunnen worden om het transport langsheen kapen en strandhoofden over zowel relatief korte (ordegrootte dagen) als langere tijdsperiodes (een jaar) te simuleren. De vergelijking met analytische en semi-empirische modellen en veldgegevens tonen aanvaardbare resultaten aan.

De resultaten van een schematisch numeriek procesmodel voor het geval van een enkelvoudig strandhoofd tonen aan dat de evolutie van de kustlijnligging en de zandtransportvolumes overeenstemmen met de uitkomst van een analytisch model. De kustlijnligging benedenstrooms van het strandhoofd wordt goed voorspeld door het numerieke model, en stemt overeen met de realiteit. De resultaten verbeteren nog wanneer ook de invloed van golfgroepen in rekening wordt gebracht. Variatie van de voornaamste invloedsfactoren voor langstransport, namelijk golfhoogte, richting en sedimentkorrelgrootte, hebben een duidelijk effect op de uiteindelijke morfologie.

De resultaten van een schematisch model met twee strandhoofden tonen aan dat de morfodynamische eigenschappen van baaien kunnen worden voorspeld met een numeriek procesmodel. Resultaten verkregen met een semi-empirisch model zijn consistent met die van het numerieke model. Muistromen in de brandingszone, in het bijzonder deze parallel aan het strandhoofd, zijn de oorzaak van zandtransport van het strand naar dieper water.

De bevindingen uit de veldstudie van een kleine baai in het oosten van Maleisië waar een strandsuppletie is uitgevoerd, tonen aan dat de seizoenaliteit in het golfklimaat een cyclisch gedrag van de strandprofielen veroorzaakt. De variaties in de dwarsprofielen langsheen de baai vertonen het klassieke patroon van zandtransport dwars op de kust. Strandplanformrotatie als gevolg van variatie in het golfklimaat in functie van de seizoenen had verlies aan zand tot gevolg bij het zuideinde van de baai. Bovendien toont de analyse van de zandvolumes van het zuidelijke deel van de baai aan dat het gesuppleerde zand uit het noordelijke deel een bijdrage levert aan het potentiële transport rond de zuidelijke kaap.

Voor de gevalstudie van de zuidelijke Gold Coast in Australië werd een numeriek procesmodel opgezet dat een permanent bypassing systeem kan nabootsen. Voorafgaand aan de morfologische modelstudie, werd het hydrodynamisch model succesvol gekalibreerd en gevalideerd aan de hand van beschikbare veldmetingen. De resultaten van het morfologische model tonen aan dat de additionele toevoer van zand via de lozingen leidt tot de vorming van een strandhaak. Het zandtransport voorbij de kaap begint onder de vorm van migratie van zandgolven van oost naar west. Deze zandgolven vormen een haakwal, die na het ronden van een tweede kaap samensmelt met het strand te westen van het landhoofd. Kwalitatief komt het morfologische gedrag van het kustsysteem zoals het wordt gesimuleerd met het numerieke procesmodel overeen met de realiteit. Het gemodelleerde bypassing-proces werkt precies zoals het wordt gepostuleerd in het conceptuele model. De slotconclusie is dat het succes van een dergelijke grootschalige permanente bypassing-operatie afhangt van het afstemmen van de tijdstippen van het lozen van grote zandvolumes op het golfklimaat.

Samengevat toont de voorliggende studie het belang aan van het correct begrip van de kustprocessen, en van de structurele geologische controle voor het inschatten van zandtransport langsheen landhoofden, en de daaraan gerelateerde morfodynamica van stranden in baaien. Het numerieke procesmodel dat in deze studie werd ontwikkeld is nuttig voor het vergroten van de kennis over de herverdeling van zand langsheen de kust onder invloed van kustwaterbouwkundige ingrepen en natuurlijke kustprocessen.

This abstract is translated from English to Dutch by Mr. Johan Reyns, Lecturer, Water Science and Engineering Department, UNESCO-IHE.

TABLE OF CONTENTS

CHAPTER 1

Introduction

This thesis explores mechanisms and processes of sand bypassing. Sand bypassing processes and embayed beaches morphodynamics in general, are a relevant but poorly understood topic. The thesis is structured from identifying the mechanisms to modelling the processes of sand bypassing in artificial and non-artificial coastal environments. This thesis ends with the summary and responses to each research question.

1.1 Background

1.1.1 Embayed beaches and sand bypassing in general

Embayed coasts, whether naturally developed or artificially created are generally associated with a dominantly closed sediment circulation system, cross-shore sediment exchange, but may also experience sediment leakage around natural headlands or artificial headland structures like groynes. Artificial headlands are developed to act as natural headland controls preventing the coast from continuous erosion. Artificial headlands can be formed from different type of coastal structures, often related to groyne structures such as I-groyne, T-groyne, and L-groyne. Headland controls can be used to stabilise straight coasts, embayed or convex coasts, reducing or even stopping littoral drift where required, but still allowing bypassing of sediment to the downdrift coast. Artificial embayed beaches, which can be created by building a series of artificial headlands along the coast have been suggested as a means of stabilising eroding shorelines (Klein et al., 2003).

Several causes of coastal erosion on embayed beaches have been reported in literature, for example, large storm events (individual or group storms), flood events, variation in wave energy, wave energy and coastal morphodynamics, longshore imbalance, and etcetera. Since erosion at those (embayed) beaches do not proceed with the high rate which is frequently having on long beaches fed by rivers, the process is probably underestimated (Pranzini and Rosas, 2007). Embayed beaches are typically affected by headland sand bypassing (Short and Masselink, 1999), by the formation of rips (Holman et al., 2006) and by beach rotation (Klein et al., 2002). Headland sand bypassing is a process of natural sediment movement from an up-coast area to a downcoast area. The occurrence of headland sand bypassing can cause readjustment of beach orientation (i.e. beach rotation) and as a result influences the long term stability of embayed beaches.

The headland sand bypassing conceptual model was first introduced by Evans (1943). The conceptual model describes the interaction between the wave-induced currents and tide-induced currents that contributes to sand bypassing around a blunt headland. Since then no other headland sand bypassing model is introduced, but studies on natural sand bypassing are still progressing, often related to the application of groyne structures on straight beaches. Pelnard-Considere (1957) was the first to develop an analytical one-line model to compute shoreline changes. The groyne sand bypassing model was also developed to calculate sand bypassing rates over the groyne tip with an assumption that the updrift groyne must be fully-filled by sand. On the basis of one-line theory, several one-dimensional analytical sand bypassing models were developed (e.g. Larson et. al, 1997; Kamphuis, 2000). All of these models are capable of calculating sand bypassing rates but none of them could potentially visualise the process of sand bypassing over the groyne tip.

Apart from the first headland sand bypassing model that was introduced by Evan (1943), another headland sand bypassing model was developed, in which the headland shape is represented by a gaussian function mimicking the real shape of a natural headland. Short and Masselink (1999) developed a conceptual model of headland sand bypassing, known as headland-attached sand bar bypassing based on their observations on several beaches in the Western Australia. Another headland sand bypassing model that was developed by Smith (2001) is recognised as a headland-strand sand bypassing. This model relates the angle between the bypassed sediment pathway and the shoreline.

Headland or coastal structures that are placed on a beach may modify beach behaviour and nearshore processes. Embayed beaches as defined earlier have different characteristics compared to open and long straight beaches. An obvious different between embayed beaches and open straight beaches is through its surf zone current circulation. On embayed beaches, the surf zone current is characterized by the presence of rip currents. Rip currents generate strong currents flowing in a seaward direction. Rip currents on embayed beaches, which can be categorised into three rip types i.e. normal beach rips, topographic rips and megarips, are responsible for the cross-shore sediment exchange on embayed beaches (Ferrari et al., 2013; Loureiro et al., 2012; Short, 1985; Coutts-Smith, 2004). The cross-shore sediment exchange is one of the mechanisms of sand bypassing.

1.1.2 Beach nourishment on embayed beaches

Beach nourishment is one of the coastal engineering techniques to provide a wider beach area for recreational purposes. Due to a closed sediment circulation system and limited sediment supply on embayed beaches, beach nourishment is often required. In Malaysia, coastal erosion may be amplified during the northeast monsoon period when high water levels, associated with high waves, result in waves breaking directly against the scarp, causing loss of materials. Though some of these materials might be returned to the shore after the monsoon, the quantity returned is normally much less; hence the net result is erosion.

The Malaysian east coasts are monsoon influenced, with erosion taking place during the northeast monsoon and accretion during the southwest monsoon, resulting in seasonal beach profile changes. Control of coastal erosion has become an important economic and social need. One of the efficient ways to address erosion problems is through a soft engineering approach. Beach nourishment is a common soft-engineering technique for beach protection in Malaysia (Ghazali, 2005), often applied to embayed beaches. The varying wave energy caused by bi-directional wave climates may modify the shape of a nourished embayed beach and may change the cross-shore beach profiles. This seasonal wave effects may also change the behaviour of flow and sediment transport patterns and thereby are the main concerns of this study.

1.1.3 The role of a permanent (fixed) sand bypassing system

Sand bypassing systems have been created to artificially bypass the littoral drift from the updrift coast to the downdrift coast in order to maintain a navigable river entrance as well as to protect neighbouring beach amenity. A number of different sand bypassing systems have been developed and employed around the world. A list of world-wide sand bypassing systems can be obtained from a report written by Boswood and Murray (2001). One of the successful sand bypassing systems is the Tweed River sand bypassing which was installed in the territory of New South Wales and Queensland, Australia. The system employs a fixed or permanent sand bypassing plant. The system has improved the navigation channel of the Tweed River and has delivered large amounts of sand to the southern Gold Coast beaches, although it has been controversial from many community perspectives (Castelle et al., 2009). The complex high energy coastal or estuarine environments with various coastal processes that lead to dynamic changes of the southern Gold Coast beaches due to this permanent sand bypassing system are under investigation of this study.

1.2 Motivation of the thesis

Sediment particles, particularly sand are naturally moving from one beach to another beach either through the alongshore or cross-shore movements. In the presence of structures (e.g. headlands or groynes) along a coast, sediment may find its own way to bypass these structures.

Sand bypassing processes play a major role in a coastal system especially in the definition of the coastal sedimentary budget. This process was found to significantly contribute to the stability of an embayed coast. However, embayed beaches research is usually focused on the beach rotation since these beaches are generally regarded as closed systems. The sand bypassing mechanisms have been extensively studied in the context of artificial structures (e.g. groynes and jetties) but studies of natural headland sand bypassing are scarce and usually applied to decadal time scales (Ribeiro et al., 2014). Recent sand bypassing studies are carried out based on large scale field experiments of tracking the sediment movement around the rocky headland (Duarte et al., 2014) and determining the bypassed sediment characteristics (Cascalho et al., 2014). Such studies which are related to the sediment bypassing processes at the tidal inlets, (e.g. FitzGerald et al., 2000; Kwok et al., 2007; FitzGerald and Pendleton, 2002) at the river inlets (e.g Tweed River project and Nerang River project) and hydrodynamics and sediment circulation patterns around large scale headlands (e.g. Bastos et al., 2003; Bastos et al., 2002; Signell,1989) might be useful to be adopted as reference cases for the present study. There are headland sand bypassing schemes presented in literatures (e.g. Evan, 1943; Smith, 2002; Short and Masselink, 1999; Storlazzi et al., 2001), but these schemes have not yet been proven. Overall, sand bypassing processes around headlands and morphodynamic of embayed beaches in general, is a relevant but poorly understood topic.

Numerical model tools are useful to understand the complex nearshore processes in embayed coasts. Long term empirical models which are derived from the hyperbolic tangent equation (Silvester, 1970), the logarithmic spiral equation (Krumblein, 1944) and the parabolic bay shape equation (Hsu and Evan, 1989) have been developed and are applicable to determine the static and dynamic equilibrium of embayed beaches. These empirical models are valid to long term empirical investigation, but not physically based. Analytical shoreline change models (Pelnard-Considere, 1956; Larson et al., 1987; Larson et al., 1997) and analytical sand bypassing models (Pelnard-Considere, 1956; Larson et al., 1987), which are derived from the one-line theory have been developed to estimate the coastline change and bypassing rate in the presence of a groyne structure. Although these analytical models have been used to compute the shoreline evolution and sand bypassing rates, yet the process of sand bypassing is still missing. The difficulty in the numerical modelling is caused by the inevitable approximations in the solution of combined refraction and diffraction in changing of water depths, non-linear wave effects, and sediment transport in wave-current flow, and etceteras. Therefore, this is one of the reasons why process-based models are highly recommended.

Numerous process based numerical models (e.g Delft3D, MIKE 21, XBeach, TELEMAC) exist in the coastal modelling world's and these models are increasingly able to solve complex coastal environment problems even if some parameters need to be adjusted in each model's configuration setting to suit its functionality. In this thesis, XBeach model was selected as a main numerical tool to model the relevant nearshore coastal processes both in hypothetical and real case studies.

In this thesis, we started developing our understanding on the topic of interest by collecting and reviewing literature materials that are related to the mechanisms and processes of sand bypassing around headlands and groynes in open straight beaches as well as embayed beaches. Based on these literature works, the mode of sand bypassing is then identified and the conceptual sand bypassing models are sketched. Further literature works are carried out to identify the main driving factors that may contribute to the natural processes of sand bypassing. Available numerical modelling tools that are related to sand bypassing phenomena and that can be used to compute the sand bypassing rates as well as being capable to demonstrate the bypassing processes are presented.

The modelling works started with the development of two hypothetical case studies i.e. (i) model cases with a single groyne structure and (ii) model cases with two groyne structures.

For the first hypothetical case study, models with a single groyne structure are built. The groyne is used to represent a non-erodible headland structure. The evolution of shorelines, sand bypassing rates and sand bypassing processes are investigated through the XBeach numerical modelling. Numerical results obtained from the XBeach model are compared with the results of analytical Pelnard and Larson models. The differences in simulated results between the XBeach model and the analytical models are evaluated.

For the second hypothetical case study, a series of models with two groyne structures is built as to represent embayed beaches. Similar to the first case study, both groyne structures are treated as non-erodible headlands. The alongshore distance between the groynes are varied so that embayed beaches can be represented by narrow and wide embayment basins. The morphodynamics of different embayed beaches are investigated through the impact of structural headlands. Embayment scaling empirical models are used to preliminary characterise the surf zone current circulation in different embayed beaches. The Xbeach model is then used to reproduce the surf zone current patterns and morphological bed patterns under the wave-group and non wave-group forcing. Further, the mechanism for transporting sediment outside the surf zone is proven through the implementation of virtual drifters in the model.

The research works are continued with the development of real case studies. Two real case studies are selected to observe the real sand bypassing phenomena in the non-artificial coastal environments. The first case study is located on the east coast of Malaysia and is characterised by a small embayed beach, influenced by a wave monsoonal effect. The beach and foreshore areas were nourished by sand to protect the beach from continuous erosion due to strong effect of wave seasonality. The spatial and temporal distributions of beach profile changes during the nourishment periods are investigated. Beach rotation and sand leakage around a headland are identified.

The southern Gold Coast beaches, located in Queensland, Australia are chosen as the second case study. Several models were built to observe the hydrodynamics and natural sand distribution patterns around the complex coastal environments involving few natural headlands (Snapper Rock, Greenmount, Kirra), a groyne structure (Kirra Point) and the Tweed River. The impact of Tweed River permanent sand bypassing system on the natural sand distribution patterns around the southern Gold Coast beaches are investigated through a numerical modelling study. An artificial sand bypassing system is introduced in a schematized way. The dynamic of swell waves may explain the characteristics of surf zone current circulations as well as sand transport patterns during calm and energetic wave conditions.

1.3 Research objectives and questions

The main purpose of this study is to "***contribute to the understanding of the mechanisms and processes of sand bypassing in artificial and non-artificial coastal environments***" through a numerical modelling study. Research objectives and research questions are formulated to achieve the main purpose of this study.

Research objectives are:
1. to identify the sand bypassing mechanisms and their relevant bypassing processes through a literature study.
2. to classify the mechanisms and processes of sand bypassing into two main categories i.e. wave-driven alongshore sand bypassing process and wave-driven cross-shore sand bypassing process.
3. to understand the alongshore sand bypassing process and the cross-shore sand bypassing process through the XBeach numerical modelling.
4. to evaluate the differences in simulated results between the XBeach model and the analytical and empirical models.
5. to investigate the morphological bed profile changes, flow and sediment transport patterns on a real embayed beach under the influence of a strong wave seasonality.
6. to investigate the impact of permanent (fixed) sand bypassing on the flow and sand distribution patterns and the morphological bed changes on a highly complex beach system.

Research questions are:
1. What are the mechanisms of sand bypassing and how the processes of sand bypassing work?
2. What are the main driving forces that may contribute to the sand bypassing process?
3. To what extent does wave seasonality affect the sand bypassing process ?
4. To what extent does the permanent sand bypassing system contribute to the natural sand distribution pattern on a complex coastal environment?
5. Are the results obtained from the process based model comparable to the results obtained in the empirical and analytical models ?
6. Do the sand bypassing processes simulated by the process based model represent the sand bypassing scenarios in reality ?

1.4 Thesis framework

The thesis contains seven chapters and the summary of each chapter is given as follows :

Chapter 1 gives a general structural framework of the overall research works. Research objectives and research questions are formulated to achieve the main purpose of the study.

Chapter 2 provides literatures of natural sand bypassing processes around natural headlands and coastal engineering structures. Two main sand bypassing mechanisms are indentified and the sand bypassing processes are thoroughly explained. Various numerical modelling approaches that are related to the sand bypassing phenomena are presented. The choice of XBeach as the numerical model program that is mainly used in the research is explicitly explained.

Chapter 3 describes the mechanism of wave-driven alongshore sand bypassing through the XBeach numerical modelling. Single groyne structure is used to represent a headland structure. The effect of wave parameters i.e wave heights and wave angle and sediment grain sizes on the shoreline evolution patterns and sand bypassing rates is investigated. Numerical results are compared between analytical and process based models. The scheme of headland sand bypassing is presented.

Chapter 4 explains partly the mechanism of wave-driven cross-shore sand bypassing through the XBeach numerical modelling. Morphodynamic investigations of embayed beaches through the impact of structural headlands are carried out. The surf zone current circulations and the morphological bed patterns in different embayed beaches are studied. Simulated results of XBeach are compared to the real phenomenon observed in the field. The hypothesis that rip currents can transport sediment out of the surf zone is proven through an implementation of virtual drifters in the XBeach model.

Chapter 5 discusses the effect of seasonal wave climates on the seasonal beach changes on a small embayed beach called *Cempedak* bay located on the east coast of Malaysia. Beach profile changes and sand volume profile changes are studied based on the analysis of field data measurement. Cross-shore sand transport and sand leakage around a small headland are both processes that influence the stability of a nourished embayed beach. A bi-directional wave climate induced by seasonal wave monsoons causes a slight rotation of the *Cempedak* bay. A conceptual model of sand bypassing across a small headland is presented.

Chapter 6 presents the impact of permanent (fixed) sand bypassing system on the natural sand distribution patterns around the southern Gold Coast beaches. Calibrated and validated wave parameter's results obtained from the models are compared against the field measurement data. An innovative model approach of an artificial sand bypassing system is introduced. Sand bank discharge operations that may affect the behavioural patterns of wave-induce currents and distribution of sand along the southern Gold Coast beaches are studied.

Chapter 7 summarizes the overall works presented during the Ph.D study. The general findings of each chapter are presented. Each research answer is addressed in response to each research question. Practical implications of the findings and recommendations are given for the future research improvement.

References

Bastos, A., Collins M., and Kenyon, N. (2003). Water and sediment movement around a coastal headland: Portland Bill, southern UK. *Ocean Dynamics* 53:309-321. DOI: 10.1007/s10236-003-0031-1.

Bastos, A.C., Kenyon, N.H., and Collins, M. (2002). Sedimentary processes, bedforms and facies, associated with a coastal headland: Portland Bill, Southern UK. *Marine Geology* 187: 235-258.

Boswood, P.K and Murray, R.J. (2001). Worldwide sand bypassing system: Data report. Conservation Technical Report No.15, Queensland Australia. ISSN 1037-4701 August 2001.

Cascalho, J., Duarte, J., Taborda, R., Ribeiro, M., Silva, A., Bosnic, I., Carapuco, M., Lira, C., and Rodrigues, A. (2014). Sediment textural selection during sub-aerial headland bypassing. An example from the Nazare coastal system (Portugal), *Jornadas de Engenharia Hidrografica*, Lisboa 24-26 de junho 2014, pp 289:292

Castelle, B., Turner, I.L., Bertin, X., and Tomlinson, R. (2009). Beach nourishments at Coolangatta Bay over the period 1987-2005: Impacts and lessons, *Coastal Engineering*, doi 10.1016/j.coastaleng.2009.05.005

Coutts-Smith (2004). The significance of megarips along an embayed coastline. Ph.D thesis. University of Sydney, Australia, 221p.

Duarte, J., Taborda, R., Ribeiro, M., Cascalho, J., Silva, A., and Bosnic, I. (2014). Evidences of sediment bypassing at Nazare headland revealed by a large scale sand tracer experiment, *Jornadas de Engenharia Hidrografica*, Lisboa 24-26 de junho 2014, pp 289:292

Evans, O. F. (1943). The relation of the action of waves and currents on headlands to the control of shore erosion by groynes. *Academy of Science for 1943*: 9-13.

Ferrari, M., Cabella, R., Berriolo, G., and Montefalcone, M. (2014). Gravel sediment bypass between contiguous littoral cells in the NW Mediterranean Sea. *Journal of Coastal Research*. 30(1): 183-191.

FitzGerald, D.M., Kraus, N.C., and Hands, E.B. (2000). Natural mechanisms of sediment bypassing at tidal inlets, US Army Corps of Engineers: USA. 10p.

FitzGerald, D.M., and Pendleton, E. (2002). Inlet formation and evolution of the sediment bypassing system: New Inlet, Cape Cod, Massachusetts. *Journal of Coastal Research* : 290-299.

Ghazali, N.H. (2006). Coastal Reclamation and Reclamation in Malaysia. *Aquatic Ecosystem Health and Management* , 9:237-247.

Holman, R., Symonds, G., Thornton, E.B., and Ranasinghe, R. (2006). Rip spacing and persistence on an embayed beach. *Journal of Geophysical Research*, 111, C01006. DOI:10.1029/2005JC002965.

Hsu, J.R.C., and Evans, C. (1989). Parabolic bay shapes and applications. *Proceedings of the Institute of Civil Engineers*. pp. 557–570.

Kamphuis, J.M. (2000). Introduction to Coastal Engineering and Management, Singapore : World Scientific Publishing, 437 pp.

Klein, A.H.F., Raabe A.L.A., and Hsu, J.R.C. (2003). Visual assessment of bayed beach stability with computer software. *Journal of Computer & Geoscience* .29:1249-1257.

Klein, A.H.F., Benedet, L., and Schumacher, D.H. (2002). Short term beach rotation processes in distinct headland bay beach system. *Journal of Coastal Research* 18:442-458.

Krumblein, W.C. (1944). Shore processes and beach characteristics. Technical Memorandum N°3, Beach Erosion Board, U.S. Army Corps Engineers (1944). 47 p.

Kwok, F.C, Gerritsen, F., and Cleveringa, J. (2007). Morphodynamics and sand bypassing at Ameland Inlet, The Netherlands. *Journal of Coastal Research* 1:106-118.

Larson, M., Hanson, H., and Kraus, N.C. (1987). Analytical solutions of the one-line model of shoreline change. *Tech. Rep. CERC-87-15,USAE-WES*, (Vicksburg, Miss.: Coast. Eng. Rest. Clr.).

Larson, M., Hanson, H., and Kraus, N.C. (1997). Analytical solutions of one-line model for shoreline change near coastal structures. *Journal of Waterway, Port, Coastal, and Ocean Engineering*, 123 (4): 180-91.

Loureiro, C. Fereira, O., and Copper, J.G. (2012). Extreme erosion on high energy embayed beaches; Influence of megarips and storm grouping, *Geomorphology*, 139-140: 155-171.

Pranzini, E., and Rosas, V. (2007). Pocket beach response to high magnitude-low frequency flood (Elba island,Italy). *Journal of Coastal Research*:969-977.

Pelnard-Considere, R. (1956). Essaidetheoriedel evolution desforms derivagesen plage desableetdegalets. *Fourth Journeldel' Hydralique, lesenergiesdela Mer, QuestionIII,*, Rapport No.1, 289–98.

Ribeiro, M., Taborda, R., Lira, C., Bizarro, A., and Oliveira, A. (2014). Headland sediment bypassing and beach rotation in a rocky coast: an example at the western Portugese coast. *Geophysical Research Abstract*, vol. 16, EGU2014-14602.

Signell, R.P. (1989). Tidal dynamics and dispersion around coastal headlands, Woods Hole Oceanographics Institution, Massachusetts Institute of Technology, USA. 162p.

Silvester, R. (1970). Development of crenulated shaped bays to equilibrium. *Journal of Waterways and Harbors Division*, ASCE, 96 (WW2) (1970), pp. 275–287

Short, A.D. (1985). Rip current type, spacing and persistence, Narrabeen Beach, Australia, *Marine geology*, 65: 47-71.

Short, A. D. and Masselink, G. Eds. (1999). Embayed and structurally controlled beaches, In: Short, A.D. (ed).,*Handbooks of Beach and Shoreface Hydrodynamics.* Chicester: John Wiley & Sons. pp.230-249

Smith, A.W. (2001). Headland bypassing. Coasts & Ports 2001: *Proceedings of the 15th Australasian Coastal and Ocean Engineering Conference*, the 8th Australasian Port and Harbour Conference, Institution of Engineers, Australia, Barton, A.C.T. (2001), pp. 214–216

Storlazzi, C.D. and Field, M.E. (2000). Sediment distribution and transport along a rocky, embayed coast: Monterey Peninsula and Carmel Bay, California. *Marine Geology*, 170(3-4): 289-316.

CHAPTER 2

Review of sand bypassing processes and modelling approaches

The sand bypassing mechanisms, either around natural headlands or engineered groyne structures classified as wave-induced alongshore bypassing and wave-induced cross-shore bypassing were identified. In the first mechanism, high-oblique waves cause the beach to rotate and accumulate sand at the updrift side of the structure. The sand moves around the structure forming a bar, which attaches on the leeside of the structure before it merges to the downdrift beach. In the second mechanism, sand must be first moved cross-shore, out past the surfzone. The cross-shore movement most likely occurs due to rip currents. Once the sand has moved offshore, it must be advected to bypass the intervening structure. A small wave height obliquely approaching the shoreline could initiate alongshore currents that would sweep the sand and deposit it in the nearshore, feeding the downcoast beaches. The modelling approaches related to the sand bypassing and embayed beaches are presented.

2.1 Introduction

Sand bypassing around headlands is one of the factors that may contribute to the stability of embayed beaches (Klein et al., 2010 and Silveira et al., 2010). Sediment transport in the vicinity of embayed beaches especially near a headland is complex, making it one of the most difficult coastal systems to be quantified in a coastal environment. Moreover, headland sand bypassing mechanism is crucial to understand a decadal scale shoreline change on open obliquely-oriented coasts. Yet, there is insufficient literature on this topic (headland sand bypassing), often related to engineering projects (Goodwin et al., 2013).

Sediment flow has been widely observed to bypass obstacles such as reefs, ebb tidal deltas, groyne, and breakwater that interrupt longshore transport at low and high wave angles. However, headland bypassing process that permit the exchange or leakage of sediment in deeply embayed compartments are largely undocumented. The presence of a large headland between two beaches disrupts wave-induced longshore transport in the surf zone, thus minimises longshore gradients. As a consequence, the alongshore transport requires a cross-shore sediment exchange and alongshore transport outside the surf zone to bypass the intervening headland. Until now, sediment transport processes at headland bay beaches are still under investigation, and further studies are always required.

Several mechanisms of natural sand bypassing have been reported in literature (e.g. Evans, 1943, Short and Masselink, 1999, Storlazzi et al., 2000, Smith, 2001, Ferrari et al., 2013), however, none of them is proven through the numerical modelling.

Scholar and Griggs (1997) for example, elucidated that there are four paths of sediment to be transported on an embayed beach. They are either moving around the headlands, transporting over rock outcrops, onshore and offshore movement or mobilizing within an isolated pocket beach. Although the existence of these pathways has been identified, little is known the philosophy that lies behind these sediment's pathways.

Understanding the basic process of headland sand bypassing is essential to improve knowledge on the stability of embayed beach systems. Therefore, it is important to first identify the existing mechanisms of natural sand bypassing around headlands observed in reality and classify them (mechanisms) based on their processes. Therefore, this chapter is devoted to describe those following specific objectives:

(i)   to describe the characteristics of headland and stability of embayed beaches;
(ii) to define the wave-induced alongshore sand bypassing and describe step-by-step the process of sand bypassing;
(iii) to define the wave-induced cross-shore sand bypassing and describe step-by-step the process of sand bypassing;
(iv) to review available modelling approaches i.e. analytical models, an empirical model and a process-based model that are related to shoreline change, sand bypassing process and long term stability of embayed beaches.
(v)   to investigate the processes of sand bypassing across a short groyne and a headland structure.

2.2 Headland and embayed beaches

2.2.1 Headland characteristics

A headland is defined as a piece of land that protrudes into a water body. It is characterised by high-breaking waves, rocky shores, intense erosion, and steep sea cliffs. It is also defined as natural promontory of relatively hard material on a sandy coastline (see **Figure 2-1**). They act as groynes and compartmentalise a coastline. One large isolated headland usually causes to form an embayment on its downdrift shoreline. The important characteristics of a headland was described by Van Rijn (1998) as :

(i)   convergence points for wave energy;
(ii) obstruction to longshore tide and wind-induced currents, convergence of currents;
(iii) large scale circulation zones downstream of headlands;
(iv) an obstruction to littoral drifts;
(v)   fixation points for seaward rip currents;

(vi) fixation points for spit formation, wing-type headlands;

(vii) fixation points for offshore shoals, originating from headland erosion.

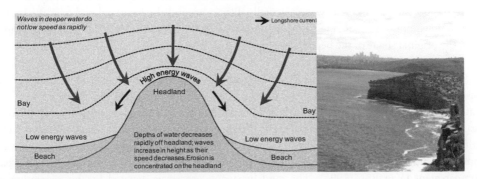

Figure 2-1. A schematised diagram of a headland, propagation of waves and longshore current (left panel). Right panel shows a photo of a real headland.

Headland behaves as a natural obstacle to alongshore sediment transport. This results in sediment deposition at the up-drift side of the headland and erosion at the down-drift side of the headland. Erosion problems in regions down-drift side of the headland are subjected to a high energy wave associated with a high angle of breaking wave along the coastline. This high breaking wave angle causes strong longshore currents and transports sediment along the shore. Instead of an orientation of the wave breaking relative to the shoreline (wave breaking angle), wave heights and sediment grain sizes may contribute to the variation of sediment transport rates.

Headlands may act as a cell boundary where local sediment circulation is confined into compartments (Bray et al., 1995). When headlands are present and widely spaced and or beach receives low waves, the beach becomes sub-intermediate, as the headlands' impact surf zone circulation only next to the headlands with a normal circulation in between (Short and Masselink, 1999).

2.2.2 Embayed beaches stability

Embayed beach usually consists of three zones, *i.e.* shadow region (zone of the lee side of up-coast headland, almost circular section behind the headland), centre region (in between the shadow zone and tangential zone- logarithmic spiral curve) and tangential region (zone of relative straight stable near the down coast headlands) (Kenneth, 1981 and Van Rijn, 1998). The stability of such embayed beaches are basically based on three states, *i.e.* static equilibrium state, dynamic equilibrium state, and natural reshaping or unstable state (Hsu and Evan, 1989; Hsu et al., 2008). These states can be identified through a modelling study by using an empirical model known as the parabolic bay shaped model (Klein et al., 2003).

In a static equilibrium state, the net longshore sediment transport is approximately zero. It also can be characterized by the presence of storm and or swell waves that come from one dominant direction, no further beach changes, and simultaneous wave breaks at every location along the beach. For this case, waves generally diffract around the headland and near the beach. In a static equilibrium state, the bay shape modelled by the parabolic bay shape model coincides with the original shape of the beach (see **Figure 2-2, left panel**)

On the other hand, beaches in a dynamic equilibrium state depend on sediment budget. It is basically differed from the static state equilibrium. The dynamic beach line lies seaward of the static beach line where the shoreline will migrate landward if sediment supply from the up-coast ceases. Littoral transport may be constant, if there is a continuous supply of sand around the up-drift headland. Two factors that may influence the curvature shape of the dynamic beach line are sand bypassing from the up-drift headland and sediment input of a river outlet within the bay compartment (Van Rijn, 1998). The sand bypassing process is dictated by the generation of longshore currents, thus, transporting sediment when the wave breaks to the shoreline. Contrast to the case of the static equilibrium state, the shape of the beach does not coincide with the bay shape that is modelled by parabolic bay shape model, as presented in **Figure 2-2, middle panel**. The shoreline tends to move toward the static equilibrium state if the sediment decreases and may cause erosion and damage infrastructures near the beach area.

Natural beach reshaping or unstable state occurs when there is a coastal engineering structure constructed or placed on the beach such as groynes and breakwaters. In this case, beach tends to be eroded at the downdrift side where sediment is transported at the sheltered area as seen in the **right panel** of **Figure 2-2**. When the shoreline displacement is significantly verified, the beach is classified as an unstable state.

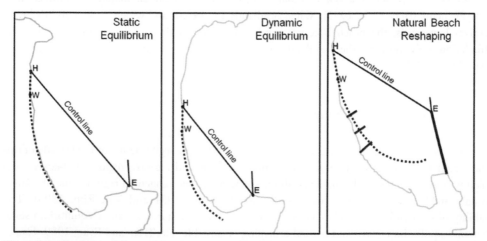

Figure 2-2. Planform stabilities of headland bay beaches. The bold dotted line indicates the static equilibrium shoreline position, described by parabolic bay shape equation. H=down-coast control point; W=down-coast tangent point; E= up-coast diffraction point. Courtesy of Ab Razak et al., (2013a).

Numerous scientific studies have been conducted considering the stability of embayed beaches. Recent study conducted by Silveira et al., (2010) for instance, investigated the relationship between the planform stability and beach characteristic such as morphodynamics state and shoreline orientation as observed by Klein et al., (2010). In the study of Silveira et al., (2010), they found no clear connection among these two factors. In their analysis, a number of 166 beaches of Southern/South Eastern Brazil were chosen to be tested and validated with the parabolic bay model i.e. the MEPBAY software. Their findings concluded that sectors with greater fluvial sediment inputs tended to have a greater number of beaches in the dynamics's state, while the most rugged sectors and fewer sediment inputs were dominated by static equilibrium state.

In a different study conducted by Jackson and Cooper (2010), the application of MEPBAY software was lead to the disagreement between the existing shoreline and the equilibrium planform stability due to difficulty in defining the wave diffraction point, disequilibrium of beach, and geological control of beach morphology. Likewise, Lausman et al., (2010) reported that there were uncertainties in the application of parabolic bay shaped model such as distortion of data that was introduced by the conversion of different software programs, the resolution of image sizes and acquisition of exact data were among the uncertainties that were recorded in their studies, despite the fact that MEPBAY model was capable to produce proper results.

2.3 Wave-induced alongshore sand bypassing

In this section, the first sand bypassing mechanism known as wave-induced alongshore sand bypassing is described. The assumption in this mechanism is that sand could easily bypass structures (headland/groyne) as the structure is considerably short that can allow continuous sand bypassing.

Bray et al., (1995) identified the 'leaky' (loss of sediment) nature of headland boundaries to littoral cells. Small headlands form "fixed partial" barriers at the boundaries to littoral cells allowing sediment to pass in both directions, although most of the small headlands only permit transport in one predominant direction. Fixed partial refers to boundary which permits the transport of sediment over it. Bypassing of small headlands appear to occur quite frequently and is associated with normal storms. Alternatively, large headlands forms 'fixed absolute" barriers at the boundaries and are much less easily bypassed, possibly except during exceptional storms which are highly intermittent (Bray et al., 1995). When large headlands protrude well offshore and into deep water, it is often assumed that there is no bypassing. However, even if there is no bypassing there still may be sediment loss from the littoral cell, as sediment may make its way along the headland shore and into the deep water. The presence of large headlands that extend out into and often past the low energy surf zone preclude the littoral transport mechanism (Storlazzi and Field, 2000).

2.3.1 Headland sand bar bypassing

In the study of Short and Masselink (1999), sand bypassing process around a headland was explained which was based on the observation of several small headlands in the northern New South Wales and south Queensland, Australia. This sand bypassing process around a small headland is illustrated in **Figure 2-3**.

Initially, the longshore transport assisted by beach rotation probably causes the sand to accumulate up-drift of the headland (**Figure 2-3,** Panel A). The sand manifests itself as a substantial sub-aqueous sand wave on the tip and immediately down-drift of the headland (**Figure 2-3,** Panel B). The sand wave which is formed when the strong tidal stream moves around and along the down-drift side as an elongated spit which often encloses the backing lagoon (**Figure 2-3,** Panel C). When the sand wave or spit attaches to the beach, it initiates a topographically controlled rip that migrates in advance of the wave, often causing severe localized beach erosion (**Figure 2-3,** Panel D). The sand finally merges with the beach, causing slight accretion (**Figure 2-3,** Panel E).

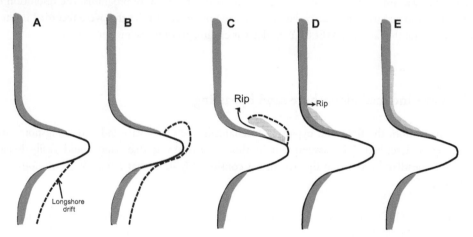

Figure 2-3. A scheme of headland sand bar bypassing (Short and Masselink 1999)

Additionally, Short and Masselink (1999) reported that the bypassing of sand on medium to high littoral drift coastlines around the headlands extending beyond the surf zone has a tendency to occur 'slugs' or pulses of sand (see **Figure 2-4**). This process is initiated by high wave energy due to larger or longer wave periods. In their study, they then added if the headland is short (such as Burleigh Heads on the Gold Coast, Australia) and the tip of the headland is in water depths shallower than closure depth, sand bypassing is more readily to occur.

Figure 2-4. Sand slug (red circle) observed in Coolangatta bay. Snapshot image was taken in October 2010. (Courtesy of Google image, 2013)

2.3.2 Headland bypassing strands

On the other hand, Smith (2001) described the concept of headland bypassing based on his observations through the headland bay beaches in Australian eastern coastline. He explained that sand is transported across the shoreface in the downdrift embayment following a strand at an angle of approximately 60^0 to the downdrift shoreline. However, there have been no coastal process or marine geological studies into this mechanism (Goodwin, 2013)

A common assumption is that most of the littoral drift flows along the coast between the wave break and the beach profile maximum uprush at high tide. However, no where within an embayment are there any wave driven hydraulic processes that could possibly be capable of executing the drastic changes of flow direction. Thus, it became apparent that an almost straight strand of sand on the seabed "jumps" the hook and ultimately carries all the bypassed sand.

The strand is initiated at the headland and lands upon the downdrift beach at, or very close to, the null point of the wave refraction orthogonal. The bypassing strand converts the zone shoreward of the strand into a closed rotating cell opposite to the strand until the transport reaches the headland control point again. Beyond this point, the cell transport contributes to the strand flow again and a rotary quasi-stability is attained (see **Figure 2-5**).

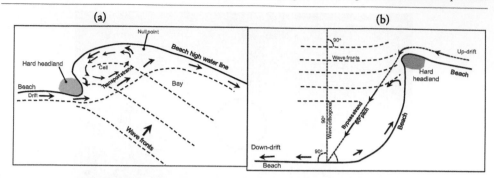

Figure 2-5. (a) Headland bypassing strand into zeta bay under oblique waves and (b) littoral bypassing strand graphical geometry for over 80% of headland bypassing systems in Eastern Australia (modified from Smith, 2001)

A pure headland bypassing strands will only occur when the shoaling seabed is sediment rich. In some cases, sand bypassing can and does occur around man-made groynes and jetty inlet. In the work of Smith (2001), he did find that fifteen out of eighteen sites experienced bypassing strands which had angles of $60^0 \pm 5^0$, maximum. The larger the distance between the headland and the downdrift beach is , the deeper the hook section of the zeta embayment will be. On the other hand, when the downdrift beach is highly curved in plan, such as many captive pocket beaches,without a straight section of beach it may be nearly impossible to plot the bypass strand and or the 90^0 wave orthogonal.

Despite the conceptual sand bypassing model proposed by Evans (1943) is not directly linked to the strand at an angle of ~60^0 to the downdrift shoreline as discussed by Smith (2001), he did find a similar mechanism of headland sand bypassing process which is driven by the forcing of wave and current (**Figure 2-6**).

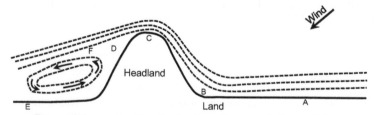

Figure 2-6. A concept of headland sediment bypassing by Evans, 1943.

Generally, currents are produced along the shoreline. As at Point **A**, the shore currents carry any sediment with it that may be present in suspension. In addition, the sand in beach ridge which forces forward by swash and backwash of the waves contributes to the movement of sediment along the shore. The beach drifting movement is slowed up where the shoreline begins to curve outward at the base of the headland until Point **B**. When the wave crests and shoreline are parallel, the forward movement ceases. The shore current (caused by tide) is also checked here and is forced to drop its energy. This results in rapid shallowing of the water and prograding

18

(move landward) of the shore. Little sediment (if any obtained along the shore at Point **A)** is carried down beyond the headland. Thus, the indentation at Point **B** is rapidly filled resulting in accretional of sand. If the wind is strong resulting in the increase of wave height, outward moving currents will be set up and some of this accumulated material at Point **B** will be carried out to the windward and distributed over the bottom of the seafloor.

Once the wind is nearly parallel with the shoreline at Point **A**, such wind-wave currents combine with shore-current and some of sediment in suspension is carried rapidly along the headland and out past its end. Also, some beach drifting will take place across the tip of the headland at Point **C**. This may result in the building of a spit to leeward. On the contrary, there is some erosion at Point **D** and Point **E.**As the waves pass the headland, they are refracted to be more nearly parallel with the shore at Point **D,** causing some beach drifting which increases towards Point **E.** If any part of the shore-current moves past the tip of the headland, it is not at once turns back along the shore but continues on for some distances before approaching it again. As a result, a reverse current is set up in the lee side of the headland. A part of the sediment carried by this reverse current may enter the main current at Point **F** or part of it may be temporarily deposited in the slack water just inshore from the place where the two currents join together. Thus, there is deposition of sediment and prograding on the lee side. Some of the materials carried beyond the outer end of the headland by the current is deposited on the bottom just to leeward. This may take the form of either ridge or a broad shelf depending on the amount of sediment available and the strength of the waves and currents. Where this deposit joins, it may also be added by the material brought by the beach drifting across the outer end of the headland. In this way, the deposit is brought above the water and extended to leeward as a spit. This spit will be broad or narrow depending on the ratio between the strength of the waves and the amount of sediment available.

2.4 Wave-induced cross-shore sand bypassing

In this sub-section, the second mechanism of sand bypassing i.e. wave-induced cross-shore sand bypassing is presented. This mechanism is directly linked to the stormy wave condition and the sand bypassing process is relatively complicated. This mechanism is often related to the presence of surf zone rip currents (hydrodynamically-controlled rips and topographically-controlled rips) in embayed beach systems.

2.4.1 Bypass of sediment from cell boundaries

Bypass of sediment from cell (headland) boundaries is likely due to a combination of cross-shore and alongshore movements as hypothesised by Storlazzi and Field (2000). In the study of Storlazzi and Field (2000), in order to transport significant volumes of sediment from one pocket beach to another, sediment must first be moved cross-shore , out past the surfzone, bypass the intervening headland (see **Figure 2-7**). This likely occurs during the passage of storms that would generate strong offshore bottom flow near the bed due to set-up caused by wave-and wind-induced onshore transport. The orbital wave motions would entrain sediment that would then be carried offshore by either seaward near-bed flows driven by wave induced setup or rip currents. Ferrari et al., (2013) reported that the bypass of coarse materials towards the adjacent littoral cell was provoked by a rip current.

Once the sediment has moved far enough offshore to bypass the intervening headland, the sediment must be advected (supension), from a much greater depth than in the surf zone, to a sufficient height to allow alongshore flows to carry the sediment over the inter-basin bathymetric highs (see **Figure 2-7,** Point **2**).

As wave heights decrease, small short-period waves obliquely approaching the shoreline could initiate alongshore currents that would sweep the sediment deposited in the nearshore, feeding the downcoast beaches (see **Figure 2-7,** Point **3**).

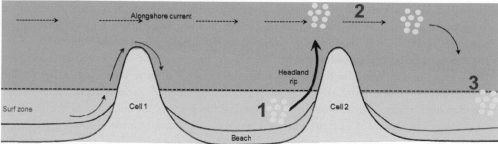

Figure 2-7. An illustration of a cross-shore sand bypassing process, sediment is transported out passes the surf zone assisted by the presence of headland rip and deposited offshore. Sediment is then transferred to the neighbouring beach by the alongshore currents.

Wright (1978) discussed the cross-shore exchanges of sediment between the beaches and the shoreface. These cross-shore exchanges are primary to the periodic changes in beach and surf zone sediment storage over intermediate time scales as well as to long term erosional and accretionary trends. This coincided with the work of Tait (1995) who pointed out that the alongshore transport of littoral sediment along rocky coasts required cross-shore exchanges and alongshore transport outside the surf zone to bypass the intervening headlands.

2.4.2 Surf zone currents in embayed beaches

Short and Masselink (1999) provided a description of headland attached rip head leakage bypassing between swash-align embayments that allows sand to be transported near the depth of closure and outside of the headland position. A distance between two headlands will affect the embayment and contributes to the formation of rip currents at the end of each headland (see **Figure 2-8**), commonly known as headland rips.

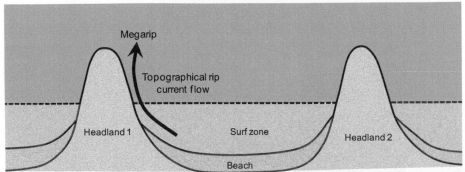

Figure 2-8. Headland structural impact on surf zone current circulation (Modified from Short and Masselink,1999)

The presence of these type of currents is identifiable through the embayment scaling parameter (δ) (see **Equation 2.1**)that was originally developed by Martens et al., (in press), but further elaborated by Short and Masselink (1999). However, assumptions made by limiting the surf zone width restrict the application of this parameter to the real life application especially for the case in which headland tip is larger than surf zone width. Based on the limitation of this parameter, Castelle and Coco (2012) recently developed δ' (see **Equation 2.2**)that can fit into embayment. The formation of surf zone currents in structurally controlled embayed beaches and detailed explanation of these formulas is given in **Section 4.1.1**.

$$\delta = \frac{S^2}{100\, C_l H_b} \tag{2.1}$$

$$\delta' = \frac{L\, \gamma \beta}{H_s} \tag{2.2}$$

2.4.3 Megarip currents

Megarip currents increase the possibility of sediment bypassing across headlands. These currents, which were first described by Short (1985) occurred due to the topographical headland control. Short (1985) explained that when the headlands are closer together and the wave height increases, the entire beach circulation may eventually become impacted by the headlands. In extreme conditions with wave height exceeding a few meter, large scale, topographically controlled cellular circulation, called megarips, prevail. **Figure 2-9 a, b, and d** shows the example of megarip currents developed in a small compartment beach. In this example, as the beach's length is relatively short, the headlands are close to each other. The megarip currents are developed adjacent to the headland or centre of the embayment, carry high velocities and can mobilize greater amount of sediment and coarser materials. As they penetrate further out to sea, they can carry this material a greater distance out from the shore.

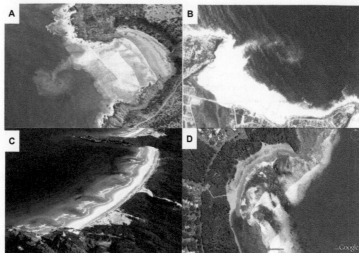

Figure 2-9. Examples of rip current development (all figures present megarip currents extending beyond the surf zones excepting panel C is normal beach rip in a transitional beach state) in embayed beaches system.

2.5 Analytical shoreline changes and analytical sand bypassing models

A qualitative understanding of basic properties of complicated physical phenomena can be obtained using analytical solutions derived from simplified problems. The analytical solution of shoreline change is often valuable and gives qualitative insights into the shoreline evolution in response to a new coastal structure or changes in environmental forcing. However, one should be aware of the limitations of these types of solutions.

Pelnard-Considere (1956) and Larson et al., (1987), for instance developed analytical formulations to estimate the shoreline evolution and sand bypassing rate at a groyne field based on the one-line theory. Their analytical solutions provided valuable qualitative and quantitative insights into the shoreline response in the vicinity of coastal structures. Additionally, LeMehaute and Brebner (1961) discussed the solutions for shoreline change at the groynes, with and without bypassing of sand, and the effect of sudden injection of sand at given point. Bakker and Edelmen (1965) modified the longshore sand transport rate equation to allow for an analytical treatment without linearization. Bakker (1969) extended the one-line theory to include two line (called two-lines theory) describing beach planform changes. The two-lines theory provided a better description of sand movement downdrift of a long groyne because it described the changes in the contours seaward of the groyne head. The two-line theory was further developed in Bakker et al., (1971), in which diffraction behind a groyne was treated. Other analytical models applications are comprehensively reviewed in Larson et al., (1997). On the other occasion, Walton Jr. (2005) reviewed inlet bypassing solutions using nomographs with the extension of the Pelnard model. Through this solution, an overall material deficit downdrift of the structure can be determined.

Recently, Walton Jr. and Dean (2011) provided a novel analytical solution that allowed the previous constant wave condition solution of Pelnard-Considere (1956) to be improved to the case where wave properties are time varying.

2.5.1 One-line theory model

A major assumption of the one-line theory is that the longshore sand transport takes place uniformly over the beach profile down to a certain limiting depth called the closure depth (D_c). Cross-shore transport is either neglected or included as a sink or source term in the sand conservation equation. The principle of mass conservation applies to the system at all times. Following the above assumption, mass conservation of sand along an infinitely small length of the shoreline can be formulated with negligible contribution from source and sink. The one-line theory model begins with the balance equation as expressed in **Equation 2.3**.

$$\frac{\partial Q}{\partial x} + D_c \frac{\partial y}{\partial t} = 0 \tag{2.3}$$

where Q=longshore transport rate (m³/s); x=space coordinate along the axis parallel to the trend of the shoreline (m); y=shoreline position (m); and t=time (s). **Equation 2.4** describes that the longshore sand transport is balanced by changes in the shoreline position. To further solve the equation above, the longshore sand transport rates, Q needs to be specified which can be defined as:

$$Q = Q_o \sin(2\alpha_b) \tag{2.4}$$

with Q_o is the amplitude of longshore sand transport and α_b = breaking wave angle. Various formulation of Q_o can be found in literature mainly based on empirical results. For this study, the formulation from the Coastal Engineering Research Centre (CERC), SPM (1984) was applied which gives the following expression:

$$Q_o = (H_b C_{gb}) \frac{K}{16\left(\frac{\rho_s}{\rho} - 1\right)(1-p)1.416^{\frac{5}{2}}} \tag{2.5}$$

where H_b = breaking wave height (m); C_{gb} =wave group velocity at breaking point (m/s); K=empirical constant (0.77); ρ= density of seawater (kg/m³); ρ_s= density of sand (kg/m³); and p= sand porosity (0.4). In addition to **Equation 2.4**, the breaking wave angle may be expressed as

$$\alpha_b = \alpha_o - \arctan\frac{\partial y}{\partial x} \tag{2.6}$$

Inserting **Equation 2.6** into **Equation 2.4**, yields

$$Q = Q_o \sin\left\{2\left[\alpha_o - \arctan\left(\frac{\partial y}{\partial x}\right)\right]\right\} \tag{2.7}$$

If it is assumed that the breaking wave angle to the shoreline and also the angle between the shoreline and x-axis is considerable small, then **Equation 2.7** can be reduced to the following expression.

$$Q = 2Q_o\left(\alpha_b - \frac{\partial y}{\partial x}\right) \tag{2.8}$$

When the amplitude of longshore sand transport rate Q_o as well as the breaking wave angle is assumed to be constant (independent to x and t), **Equation 2.3** and **Equation 2.8** can be rewritten as:

$$\frac{\partial y}{\partial t} = a\frac{\partial^2 y}{\partial x^2} \tag{2.9}$$

and a (m²/s) is a diffusion coefficient which can be expressed as **Equation 2.10**. **Equation 2.9** is identical to the one dimensional heat diffusion equation. This relationship can be used to determine the solution changes but with an appropriate boundary conditions.

$$a = \frac{2Q_o}{D_c} \tag{2.10}$$

2.5.2   Pelnard-Considere model

The first well known solution to a shoreline in the vicinity of a coastal structure is found by Pelnard-Considere (1956) in which an assumed complete blockage of the longshore littoral transport leads to the following shoreline planform solution updrift of the structure:

$$y(x,t) = \alpha_b\left(\frac{2\sqrt{at}}{\sqrt{\pi}}e^{\left(\frac{x}{2\sqrt{at}}\right)^2} - x.erfc\left(\frac{x}{2\sqrt{at}}\right)\right) \qquad t < t_{fill} \tag{2.11}$$

with α_b = breaking wave angle (radian); a = shoreline diffusivity (m²/s); x = cross-shore distance (m) and t = instantaneous time (s). **Equation 2.11** is valid until the groyne is filled by the sand and the time when the accretion reaches the tip of the groyne can be estimated as:

$$t_{fill} = \frac{\pi}{4a} \cdot \frac{L_g^2}{\alpha_b^2}$$

(2.12)

where L_g is the groyne length. For a fixed wave climate, **Equation 2.12** reveals that the time required for the shoreline to reach the end of the groyne will increase fourfold if the groyne length (L_g) is doubled (Larson et al., 1987). If bypassing of a groyne occurs at y(0,t)= L_g, **Equation 2.11** is no longer valid and it reduces to the following relationship:

$$y(x,t) = L_g \cdot erfc\left(\frac{x}{2\sqrt{at_{bypass}}}\right) \qquad t > t_{fill}$$

(2.13)

where t_{bypass} is calculated as $t_{bypass} = t - 0.38t_{fill}$. Accordingly, the rate of sand bypassing around the groyne is computed based on sand transport rate to produce the following relationship in which Q_o represents the amplitude of longshore sand transport and can be derived from various formulations *e.g* Coastal Engineering Research Centre (CERC) and Kamphuis (1991) :

$$Q_{bypass} = 2Q_o\alpha_o\left(1 - \frac{L_g}{\alpha_o\sqrt{\pi a t_{bypass}}}\right)$$

(2.14)

2.5.3 Larson model

On the other hand, Larson et al., (1987) presented an alternative shoreline change and bypassing solution at a groyne or jetty. His solution allowed a continuous shoreline evolution but assumed that sand bypassing started immediately after construction of a groyne. The beach planform solution is calculated as follows:

$$y(x,t) = L.erfc\left(\frac{x}{2\sqrt{at}}\right) - L.e^{\alpha_b\frac{x}{Lg}+\alpha_b^2\left(\frac{at}{Lg^2}\right)}.erfc\left(\alpha_b\frac{\sqrt{at}}{Lg} + \frac{x}{2\sqrt{at}}\right)$$

(2.15)

and the sand bypassing rate is computed by **Equation 2.16**. It would be expected that the shoreline evolution updrift of the groyne is slower than the Pelnard method due to the assumption of immediate sand bypassing after construction of a groyne.

$$Q_{bypass} = 2Q_o\alpha_o\frac{y}{L_g}$$

(2.16)

25

2.6 Parabolic bay shaped empirical model

Long term empirical models for a headland bay beach have been identified. They are based on empirical equations known as the logarithmic spiral equation, hyperbolic tangent equation, and parabolic bay shaped equation. Despite the fact that these empirical models have been long time existed, they are incapable to predict the rate of sand bypassing and thus impossible in describing the sand bypassing process. These models is merely applicable to determine the final shape of an embayed coastline. A chronology of historical development of these empirical models was given by Oliveira and Barreiro (2010).

One of the well-known headland bay beach formulations is recognized as a parabolic bay shaped equation (Hsu and Evans, 1989). This equation incorporates geometry of the bay and shore-face dynamics (*i.e.* diffraction points and the prevailing wave directions) as direct parameters, which are not applied in any headland bay models (see **Figure 2-10**). The parabolic bay shape equation for a headland bay beach in a static equilibrium is presented as follows:-

$$\frac{R}{R_\beta} = C_o + C_1\left(\frac{\beta}{\theta_n}\right) + C_2\left(\frac{\beta}{\theta_n}\right)^2 \tag{2.17}$$

where the geometric parameters R, R_β, β, and θ are shown in **Figure 2-10**. The constant value of C_o, C_1, C_2 is generated by regression analysis to fit the peripheries of the 27 prototypes and model bays, differ with reference angle β and can be expressed by fourth-order polynomials as written in **Equation 2.18**. These constant values can be recognized in the Coastal Engineering Manual of CERC, 2002 for longshore sediment transport in terms of graphical representation, as well.

$$C_o = 0.0707 - 0.0047\ \beta + 0.000349\ \beta^2 - 0.00000875\ \beta^3 + 0.0000000476\ 5\beta^4$$
$$C_1 = 0.9536 - 0.0078\ \beta + 0.00004879\ \beta^2 - 0.0000182\ \beta^3 + 0.000001281\ \beta^4$$
$$C_2 = 0.0214 - 0.0078\ \beta + 0.0003004\ \beta^2 - 0.00001183\ \beta^3 + 0.0000000934\ 3\beta^4$$

$$\tag{2.18}$$

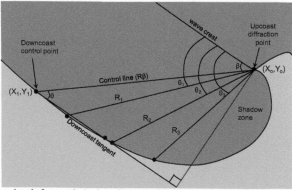

Figure 2-10. Definition sketch for parabolic bay shape model showing major physical parameters (courtesy of Hsu and Evans, 1989).

In order to obtain the two parameters (R and R_β) as in expression above, maps, topographical charts, vertical aerial photographs or satellite images of bay beaches may be used. Based on observations for static equilibrium bay beach, usual procedure for applying the parabolic model is described as follows:

1. Choose a control line with length R_β;
2. Determine the predominant wave direction and reference wave angle β;
3. Compute ray length R_β to the beach;
4. Sketch the shoreline planform in static equilibrium.

Based on the parabolic bay equation, two models have been developed which are known as Model of Equilibrium Bay Beach (MEPBAY) and Coastal Modelling System (SMC). The general purpose of developing MEPBAY software is to provide a user a friendly environment on applying the parabolic model as well as to help user arrive at an optimum design from several different options. The practical application (e.g. Klein et al., 2003; Raabe et al., 2010; Oliveira and Barreiro, 2010; Jackson and Cooper, 2010) of this software is enormous and its outputs are convincing. The operation of MEPBAY and the procedure outlined can be found in Klein et al., (2003).

Alternatively, SMC is a part of the Spanish Beach Nourishment Manual (SBM). Known as the coastal modelling system, this model is structured into five modules, *i.e.* (i) a pre-process module; (ii) short term module; (iii) middle and long term module; (iv) bathymetry renovation module; and (v) the tutorial module. The SMC also includes a graphical interface module (see **Figure 2-11**) to test the stability or to design new equilibrium beaches taking into account the equilibrium profile and planform formulation which incorporates the parabolic bay equation proposed by Hsu and Evans (1989).

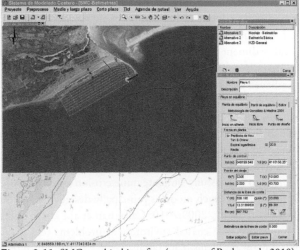

Figure 2-11. SMC graphical interface (courtesy of Raabe et al., 2010).

While SMC model supports the design phase due to the increasing forecast precision in numerical method, the MEPBAY model does not provide accurate data for this stage (design phase). Data has to be extracted from the image and the quality of the output depends directly on the image resolution (Raabe et al., 2010). Despite SMC model is accurate in design phase, it needs additional input data to allow deeper analysis of the chosen alternatives.

2.7 Process based models

Various models which are based on analytical and empirical approaches have been presented. However, these approaches have not been the most efficient way to produce the reliable long-term predictions of embayed beaches. Analytical models although applicable in predicting the shoreline evolution and sand bypassing rate, are incapable of explaining the sand bypassing process. Occasionally, results of analytical models could be under or over estimated due to the simplifications made in analytical approximations. In addition to that, analytical models provide inaccurate predictions of non-uniform processes if compared to process based models. Likewise, empirical bay models help engineers to estimate the coastline position and thus, determine the stability of embayed beaches. However, the main process that determines the evolution of the bay itself is still misleading.

2.7.1 XBeach model

Due to some above reasons, process based models like Delft3D and XBeach are needed where separation between longshore and cross-shore scales is not possible, for instance, in the vicinity of a tidal inlet, with a complex bathymetry with channels and shoals at varying angles with respect to the undisturbed coast orientation (Roelvink and Reniers, 2010). In a complex phenomenon such as sand bypassing process around a headland or an engineered structure, a two-dimensional process based model is required to resolve the longshore and cross-shore components of horizontal current and sediment transport. In this study, the XBeach model was used as a numerical tool to investigate the shoreline pattern, sand bypassing process, and the morphodynamic of embayed beaches.

XBeach, a two-dimensional process based model was developed for nearshore morphodynamics, which focuses on extreme events such as hurricanes with processes like over-washing and breaching also included. The significant reason of the innovation of this model is that both longshore variation and infra-gravity wave motions (wave groups) are found to be important factors in order to successfully simulate dune erosion and overwash in a broad range of cases. Despite the fact that the XBeach model is principally designed to predict dune erosion due to storm impact (e.g Roelvink et al., 2009:2010; Callaghan et al., 2013; Splinter et al., 2013), it also can be applied for small scale coastal environment problems (e.g Ab Razak et al., 2013b) and other fluvial environment problems (e.g Hartanto et al., 2010). Until recently, the model has been applied in various applications of beach systems. Therefore, it is worth to test the capability of this model in the application of embayed beach system by improving some model parameters, especially on the physical processes of sediment transport around coastal structures. For detailed model formulation and descriptions, readers are recommended to refer to Roelvink et al., (2009).

2.7.2 The choice of XBeach

To ensure that the sand bypassing is represented effectively, it is important that a suitable process based model be identified. The main requirement of the model is that it has to be able to effectively simulate alongshore and cross-shore sediment transport processes as well as morphodynamics in the swash region. Despite the fact that others sophisticated models are existed and can genuinely handle those processes, it is decided that XBeach is the most suitable process based model for the following reasons:

(i)   Xbeach provides a 2DH description of the short wave group, which is important for the nearshore coastal processes such as rip morphodynamics.

(ii) The implementation of stationary wave solver in XBeach which exclude some irrelevant processes makes model computations become faster. This wave solver is quit fast compare to SWAN and has no disturbance at the lateral boundaries.

(iii) XBeach takes into account the onshore and offshore transport processes which is important for the cross-shore sediment exchange.

(iv) Additionally the robustness of the model has been extensively tested through validations at different field sites.

(v)   For the long term simulations, the computational efficiency is essentially required and the fact that XBeach model can be run on multiple processors using a MPICH approach. This can save a lot of computational time.

(vi) XBeach is an open source model with an active user and support community, meaning that assistance is handy.

2.8   Model selection: Gaussian shaped or rectangular shaped headland?

2.8.1 Model setups

Prior to a comprehensive investigation of sand bypassing phenomena, a selection of a proper structure's shape should be done. This is important to understand the nature of sand bypassing process in order to simplify the investigation in our hypothetical case studies as mainly discussed in **Chapter 3** and **Chapter 4**. Two models were developed: (i) with a gaussian shape and (ii) with a rectangular shape in order to represent a natural headland and a groyne, respectively. The geometry of the gaussian shape are defined according to Signell and Harris (2000) as expressed in **Equation 2.19** and graphically presented in **Figure 2-12**.

$$f(x) = a \exp\left[-\frac{1}{2}\left(\frac{x}{b}\right)^2\right] \tag{2.19}$$

where:

$f(x)$ = coastline position (m)

b = alongshore extent (in this case 100 m)

a = offshore extent (in this case 350 m)

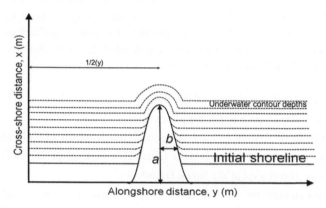

Figure 2-12. Schematization of a hypothetical model with a gaussian shaped headland.

The length of the structure was 200 m long (from the initial shoreline) protruding into a water body. A 150 m width of undeveloped beach was designed to allow for any morphological change , especially at the downdrift side of the coast. Both structures i.e gaussian and rectangular were placed in the middle of the southern boundary. For the rectangular shape, the structure's width is 10 m wide. The height of the structures is high enough i.e 8 m to prevent sand from over passing the structure's body. Bathymetric profile is linear along the cross-shore direction (beach slope of 1:40) with a uniform sediment grain size, D_{50} of 250 μm distributed elsewhere in the model domain. In front of the Gaussian shaped structure, the initial seabed contours are slightly curved following the blunt shape of the headland. The headland tip is located in a water depth of 5.4 m. An oblique wave (30^0 with respect to shore normal) was set at the offshore boundary at a water depth of 15 m propagated to the nearshore zone with the wave height of 1 m and the period of 8 s. The morphological factor of 100 was used to speed up the computational time. Models were run for at least 180 morphological days in order to allow sediment to bypass the structure. The model domain covers 1 km and 4 km in cross-shore and alongshore, respectively. A default value of 0.1 m^2s^{-1} was used for the horizontal viscosity. The model geometries used in this analysis are not meant to represent any real headland or groyne structures in reality.

2.8.2 Results of two different structural headland shapes

Figure 2-13 shows the sand bypassing process around two different structures; upper panels represents a groyne structure and lower panels represents a gaussian headland. Hypothetically, it can be concluded that both structures shows similar mechanism of sand bypassing. In both cases, sand initially accumulated at the updrift side of the structure. When the updrift structure was filled with sand, it then moved around the structure's tip, attached to the lee side of the structure and finally merged to the downdrift beach. Prior to bypassing, both structures indicated the beach area downdrift side of the structures were severely eroded.

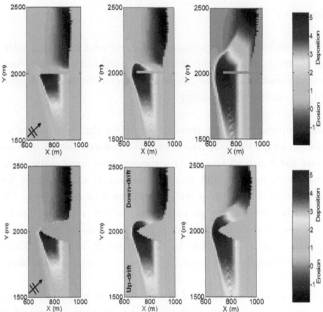

Figure 2-13. Evolution of sand bypassing processes across a short groyne (upper panels) and a gaussian shaped (lower panels) headland structure. Black inclined arrows indicate the incoming wave. The alongshore scale and cross-shore distance is cropped between 1500 to 2500 m and 600 to 1000m, respectively.

Both structural shapes are similar in their effects, causing deposition at the updrift and erosion at the downdrift side of the structure. At the southern Gold Coast Australia, a solid groyne known as Kirra Point was constructed to block the prevailing drift from north to south due to supply of sand from the Tweed River sand bypassing system. The construction of Kirra groyne is an attempt to provide a wider beach area at the Coolangatta beach. Smith (2001) showed the similarity between sand bypassing process across a groyne and a natural headland, in particular the angle between the landing strand and the downdrift beach (see **section 2.3.2**). Likewise, Short and Masselink (1999) found that the bypassing of sand around headlands or large groynes has a tendency to occurs in slugs or pulses of sand (see **section 2.3.1**). Similarly, Evans (1943) showed sand bypassing across groyne had a similar effect to sand bypassing across a natural headland. This brings about changes which finally result in the elimination of the headland. Groyne is broadly applied as an artificial headland control to combat an erosion problem in straight open coasts (e.g. Silvester and Hsu, 1993:1997). Yet, two well-known analytical bypassing models i.e. Pelnard-Considere and Larson have been developed with the application of a groyne structure which latter can be used to validate results produced by the numerical process based model. Due to the similarity between the sand bypassing process across a short groyne and a gaussian shaped headland structure, we decided to choose the groyne structure to be represented in models of our hypothetical case studies.

2.9 Conclusions

In this chapter, the mechanism and processes of natural sand bypassing were presented. Two main mechanisms of sand bypassing either through a natural headland structure or an engineered groyne structure were classified into two categories i.e. wave-induced alongshore sand bypassing and wave-induced cross-shore sand bypassing.

In general, wave components are the main driving force that contributes to the initiation of sand bypassing process. In the first mechanism i.e. wave-induced alongshore sand bypassing, high oblique waves cause the beach to rotate. As a consequence, sand starts to accumulate at the updrift side of the structure. The sand then moves around the structure forming a bar and attaches to the leeside structure before it merges with the downdrift beach. This headland sand bar bypassing mechanism is common to operate in shallow curvature planfroms. In a deep embayment, sand is transported across the shoreface in the downdrift embayment following a strand at an angle of approximately 60^0 to the downdrift shoreline. This headland strand bypassing mechanism is doubtful as no coastal or marine geological studies are done.

The second mechanism i.e. wave-induced cross-shore sand bypassing explains that the cross-shore exchange of the sand occurs due to the wave that propagates perpendicular to the shore. This mechanism brings the sand from the coast to offshore and requires cross-shore transport agents such as rip currents. As the wave height decreases, sand is most likely transported to the adjacent beach by the alongshore current, feeding the downcoast beaches. Commonly related with the generation of surf zone rip currents (normal beach rips, headland rips, or megarips), this cross-shore sand bypassing requires a special treatment for the morphodynamic modelling of embayed beaches.

Available modelling approaches related to the sand bypassing and embayed beaches were presented. The analytical Pelnard and Larson models are capable to predict the shoreline pattern and estimate the bypassing rate. The empirical embayed model was identified and its limitation was explicitly explained. On the other hand, the use of XBeach process based model as the main numerical tool applied in this study was discussed.

References

Ab Razak, M.S., Roelvink, D., and Reyns, J. (2013a). Beach response due to beach nourishment on the east coast of Malaysia, *Proceeding of the ICE-Maritime Engineering Journal*, 66(4):151-174.

Ab Razak, M.S., Dastgheib, A., and Roelvink, D. (2013b). Sand bypassing and shoreline evolution near coastal structure comparing analytical solution and XBeach numerical modelling, *Journal of Coastal Research*, SI 65: 2038-2044.

Bray, M.J. Carter, D.J. and Hooke, J.M. (1995). Littoral cell definition and budgets for Central Southern England, *Journal of Coastal Research*, 11(2): 381-400

Bakker, W. T., and Edelman, T. (1965). The coastline of river deltas. *Proceeding of the 9th Coastal Engineering Conference*, American Society of Civil Engineers, New York, 199-218.

Bakker, W. T. (1969). The dynamics of a coast with a groyne system, *Proceeding of the 11th Coastal Engineering Conference*, American Society of Civil Engineers, New York, 492-517.

Bakker, W. T., Klein-Breteler, E. H. J., and Roos, A. (1971). The dynamics of a coast with a groyne system, *Proceeding of the 12th Coastal Engineering Conference*, American Society of Civil Engineers, New York, 1001-1020.

Castelle, B. and Coco, G. (2012). The morphodynamics of rip channels on embayed beaches, *Continental Shelf Research*, 43: 10-23.

Callaghan, D.P., Ranasinghe, R., and Roelvink, D. (2013). Probabilistic estimation of storm erosion using analytical, semi-empirical, and process based storm erosion models. *Coastal Engineering*, 82: 64-75.

Evans, O. F. (1943). The relation of the action of waves and currents on headlands to the control of shore erosion by groynes. *Academy of Science for 1943*: 9-13.

Ferrari, M., Cabella, R., Berriolo, G., and Montefalcone, M. (2013). Gravel sediment bypass between contiguous littoral cells in the NW Mediterranean Sea, *Journal of Coastal Research*, 00(0), 000-000. Coconut Creek(Florida), ISSN 0749-0208.

Goodwin, I.D., Freeman, R., and Blackmore, K. (2013). An insight into headland sand bypassing and wave climate variability from shoreface bathymetric change at Byron Bay, New South Wales, Australia. *Marine Geology*, 341:29-45

Hsu, J.R.C., and Evans C. (1989). Parabolic bay shapes and applications. *Proceedings of the Institute of Civil Engineers*, Part 2, 87(4): 557–570.

Hsu, J.R.C., Benedet, L., Klein, A.H.F., Raabe, A.L.A., Tsai, C.P., and Hsu, T.W. (2008). Appreciation of static bay beach concept for coastal management and protection. *Journal of Coastal Research*, 24(1): 812-835.

Hartanto, I.M., Beevers, L., Popescu, I., and Wright, N.G. (2011). Application of a coastal modelling code in fluvial environments. *Environmental Modelling & Software*, 26: 1685-1695

Jackson, D.W.T., and Cooper, J.A.G. (2010). Application of the equilibrium planform concept to natural beaches in Northern Ireland. *Coastal Engineering*, 57: 112-123.

Kamphuis, J.W. (1991). Alongshore sediment transport rate. *Journal of Waterways, Port, Coastal and Ocean Engineering*, 117(6):624-641.

Klein, A.H.F., Raabe, A.L.A., and Hsu, J.R.C. (2003). Visual assessment of bayed beach stability with computer software. *Journal of Computer & Geoscience*, 29: 1249-1257.

Klein, A.H.F., Ferreira, Ó., Dias, J.M.A., Tessler, M.G., Silveira, L.F., Benedet, L., de Menezes, J.T., and de Abreu, J.G.N. (2010). Morphodynamics of structurally controlled headland-bay beaches in southeastern Brazil: A review. *Coastal Engineering*, 57: 98-111.

Kenneth F. (1981). Morphology variations and sediment transport in crenulated bay beaches, Kodiak Island, Alaska. *Marine Geology*, 47: 261-281.

Larson, M., Hanson, H., and Kraus, N.C. (1987). Analytical solutions of the one-line model of shoreline change. *Tech. Rep. CERC-87-15, USAE-WES*, (Vicksburg, Miss.: Coast. Eng. Rest. Clr.).

LeMehaute, B., and Brebner, A. (1961). An introduction to coastal morphology and littoral processes. (Rep. No. 14), Civil Eng. Dept., Queens Univ. at Kingston, ant., Canada.

Larson, M., Hanson, H., and Kraus, N.C. (1997). Analytical solutions of one-line model for shoreline change near coastal structures. *Journal of Waterway, Port, Coastal, and Ocean Engineering*, 123 (4): 180-91.

Lausman, R., Klein, A.H.F., and Stive, M.J.F. (2010). Uncertainty in the application of the Parabolic Bay Shape Equation: Part 1. *Coastal Engineering*, 57: 132-141.

Oliveira, F.S.B.F., and Barreiro, O.M. (2010). Application of empirical models to bay-shaped beaches in Portugal. *Coastal Engineering*, 57: 124-131.

Pelnard-Considere, R. (1956). Essaidetheoriedel evolution desforms derivagesen plage desableetdegalets. *Fourth Journeldel' Hydralique, lesenergiesdela Mer, Question III,,* Rapport No.1, 289–98.

Roelvink J.A., and Reniers, A.J.H.M. (2010). Coastal morphodynamics modelling. Singapore: World Scientific Book. 274p.

Roelvink, D., Reniers, A.J.H.M., van Dongeren, A., van Thiel de Vries, J., McCall., R., and Lescinski, J. (2009). Modelling storm impacts on beaches, dunes and barrier islands. *Coastal Engineering*, 56 (11-12): 1133-52.

Roelvink, J.A., Reniers, A.J.H.M, Dongeren, A., Vries, J.T., Lescinski, J., and McCall, R. (2010). XBeach model description and manual. (UNESCO-IHE Institute for Water Education, Deltares, Delft University Technology).

Raabe, A.L.A., Klein, A.H.F., González, M., and Medina, R. (2010). MEPBAY & SMC: Software tools to support different operational level of headland-bay beach in coastal engineering projects. *Journal of Coastal Research*, 57:213-226.

Scholar, D.C., and Griggs G.B. (1997). Pocket beaches of California. Sediment transport along a rocky coastline, in : Ewing, L., and Sherman, D., Proceedings of the California's Coastal Natural Hazards, California Shore and Beach Preservation Association, Santa Barbara, California, pp. 65-75.

Splinter, K.D., Carley, J.T., Golshani, A., and Tomlinson, R. (2013). A relationship to describe the cumulative impact of storm clusters on beach erosion. *Coastal Engineering*, 83: 49-55.

Signell, R.P., and Harris, C.K. (2000). Modeling sand bank formation around tidal headlands, in: M. L.Spaulding and A. F.Blumberg (Eds.), *6th International Conference of Estuarine and Coastal Modeling*, ASCE, New Orleans, USA.

Silveira, L.F., Klein, A.F.H., and Tessler, M.G. (2010). Headland bay beach planform stability of Santa Catarina State and of the Northern Coast of Sao Paulo State. *Brazilian Journal of Oceanography*, 58():101-122.

Silvester, R. and Hsu, J.R.C., (1993). Coastal Stabilization: Innovative concepts. Prentice-Hall, Englewood Cliffs, New Yersey: USA, 608p.

Silvester, R. and Hsu, J.R.C., (1997). Coastal Stabilization. Singapore: World Scientific, 596p-.

SPM (1984). 2 vols.(4th Ed.; Washington, D.C.: USAE-WES Coast. Eng. Res. Ctr., U.S. Govt. Printing Ofc.)

Short, A.D. (1985). Rip current type, spacing and persistence, Narrabeen Beach, Australia, *Marine geology*, 65: 47-71.

Short, A.D. and Masselink, G. (1999). Embayed and structurally controlled beaches, In:Short, A.D.(ed)., *Handbooks of Beach and Shoreface Hydrodynamics*. Chicester: John Wiley & Sons. pp.230-249

Smith, A.W. (2001). Headland bypassing. Coasts & Ports 2001: *Proceedings of the 15th Australasian Coastal and Ocean Engineering Conference*, the 8th Australasian Port and Harbour Conference, Institution of Engineers, Australia, Barton, A.C.T. (2001), pp. 214–216

Storlazzi, C.D. and Field, M.E. (2000). Sediment distribution and transport along a rocky, embayed coast: Monterey Peninsula and Carmel Bay, California. *Marine Geology*, 170(3-4): 289-316.

Tait, J.F., (1995). Rocky coasts and inverse methods: sediment transport and sedimentation patterns of Monterey Bay National Marine Sanctuary. PhD thesis, University of California at Santa Cruz, California, 138 pp.

Van Rijn, L.C. (1998). Principle of Coastal Morphology. Aqua Publications, Amsterdam, the Netherlands. 730p.

Wright, L., Thom, B., and Chappell, J. (1978). Morphodynamics variability of high-energy beaches. Coastal Engineering Proceedings, 1(16). pp 1180-1194. DOI:http://dx.doi.org/10.9753/icce.v16.%p.

Walton Jr, T.L. (2005). A review of inlet bypassing solutions with nomographs. *Coastal Engineering*, 52: 1127-1132.

Walton Jr, T.L. and Dean, R.G. (2011). Shoreline change at an infinite jetty for wave time series. *Continental Shelf Research*, 21: 1474-80.

CHAPTER 3

Natural processes of sand bypassing around a groyne structure

An investigation of natural sand bypassing processes due to a single groyne structure was carried out. The effects of wave parameters such as wave heights, wave angles and sediment grain sizes on the shoreline change and bypassing volume were assessed. A comparison between the process based numerical model and analytical model in terms of shoreline patterns and bypassing rates was completed. Some important bypassing parameters that caused differences between the results of numerical and analytical models were discussed. The natural processes of sand bypassing around the groyne structure that were captured by the XBeach model were explained. The results indicated that the characteristics of waves and sediments played an important role in determining the succession of sand bypassing process. An increase in wave heights resulted in increasing bypassing times and bypassing rates. Coarser sediment caused a slow migration of the shoreline pattern and resulted in the slow initiation of bypassing time. A shoreline comparison between the analytical Pelnard model and the XBeach model showed a good qualitative agreement between those models. However, the bypassing volume computed by XBeach was always lower than the Pelnard model. Unlike the analytical models, the XBeach model is capable of demonstrating two-dimensional processes of sand bypassing around a groyne tip. Two important parameters used in the analytical model i.e. longshore sediment transport and closure depth were found to be sensitive. Such sensitivities may cause discrepancies between numerical and analytical bypassing results. In all cases presented in this chapter, the results showed that regardless of different wave parameters and different sizes of sediment particles, a similar process of sand bypassing occurred.

Parts of this chapter were published in:

(i) Ab Razak, M.S., Dastgheib, A., Roelvink, D. (2013). Sand bypassing and shoreline evolution near coastal structure, comparing analytical solution and XBeach numerical modelling, *Journal of Coastal Research*, SI 65: 2083-2088.

(ii) Ab Razak, M.S., Dastgheib, A., Roelvink, D. (2013). An investigation of sand bypassing parameters and bypassing solution of a groyne structure, *Proceeding of the International Conference of Coast and Port 2013*, Sydney Australia, pp 1- 6.

3.1 Introduction

Wave-induced alongshore sediment promotes sand bypassing around headlands. Evans (1943) and Short and Masselink (1999) were amongst the earliest researchers to discover the mechanism of sand bypassing around headlands. Their studies were generally based on field observations and, to this day, the mechanism has never been proven through numerical modelling. The use of various numerical modelling techniques in embayed beach environments is limited to particular research efforts such as: the study of morphological development of hypothetical embayed beaches due to changing wave climates (Daly et al., 2011); determination of planform beach stability (Silveira et al., 2010; Klein et al., 2010; Jackson and Cooper, 2010; Raabe et al., 2010); and improvement of shoreline planform in crenulated bay beaches (Weesakul et al., 2010). Consequently, sand bypassing studies are only applicable to specific scenarios often related to engineering structures like groynes, offshore breakwaters, and jetties.

The computation of sand bypassing rates was first completed using an analytical approach based on the one-line theory (Pelnard-Considere, 1956). In the analytical model of Pelnard, a prototype groyne is used as a barrier to block sediment transport along the shore. Groynes have been widely used to protect beaches from severe erosion. Built perpendicular to the shoreline, groynes are specially designed to induce accretion within the groyne field and possibly to shift the coastline seaward. As a result of the raised seabed, incident waves will break further seaward. The zone of maximum wave energy dissipation is shifted so that a better protection of the coastline is achieved.

The analytical Pelnard model does not demonstrate a step-by-step process of bypassing sediment around the groyne. Nevertheless it can be used to calculate the total amount of sand that bypasses the groyne. Moreover, the model of Pelnard can estimate an ideal time (t_{bypass}) when the sand should begin to bypass the groyne tip. This estimate is based on several practical assumptions which are the incoming wave is dominantly in one direction only and the cross-shore profile is always uniform. In addition, other analytical works related to shoreline response due to installation of groyne structures such as Le Mehaute and Brebner (1961), Bakker and Edelmen (1965), Bakker (1969), Bakker et al., (1971), and Larson et al., (1987:1997) are based on the principal of one-line theory.

One of the alternate ways to understand the headland sediment bypassing process is to study sand bypassing around a hypothetical groyne structure. As the analytical approaches related to groynes have been improved, it is sufficient to use them as guidance for numerical modelling work. Sand bypassing around groynes is an important topic of interest. A groyne's design elements, including length and height, should be determined such that they allow sufficient bypassing of sand over the groyne body (over-passing), under the groyne body (under-passing), or over the groyne tip (end-passing). Nonetheless, the geometry of the groyne structure is not a focal point of this study. The main objective is to investigate how changing the wave parameters and sediment properties may influence the process of sand bypassing around the groyne.

In this chapter, the mechanism of natural sand bypassing around a groyne, previously defined as wave-induced alongshore sand transport (refer to **Chapter 2**, **Section 2.3**) is proven through numerical XBeach modelling. Leading forces such as wave heights, wave angles and sediment properties are important discussion topics related to the succession of sand bypassing process around the groyne.

Groyne is one of several engineering solutions for protection beaches from continuous erosion. Design considerations that allow sand to bypass the groyne should be addressed carefully in order to prevent localised erosion on the downdrift side of the structure. Key parameters such as wave heights, wave angles, and sediment sizes should be modelled in order to observe the behaviour of sand bypassing around the groyne structure. In longshore transport calculation formulas such, those parameters are interrelated and fundamentally important for determining the sand-bypassing rate. Therefore, specific objectives of this chapter are to:

(i) develop a simple morphological model with a single groyne structure that is used to obstruct the alongshore sediment;

(ii) investigate the effects of changing wave parameters and sediment properties on determination of sand-bypassing process and the sand-bypassing rate;

(iii) explain the step by step process of sand bypassing around a groyne;

(iv) compare the shoreline patterns and sand bypassing rates between the XBeach model and analytical models; and

(v) discuss important parameters that may influence the sand bypassing process.

3.2 Model setup

3.2.1 Grid and bathymetry

The XBeach model was applied in two-dimensional mode, in order to focus on horizontal circulations and their effects on the coastal evolution due to anthropogenic measures. This also facilitated investigation of the process of natural sand bypassing around groyne. The seabed profile had an initial slope of 1:43 and the profile shape was linear over 600 m offshore as presented in **Figure 3-1(b)**. A 150 m width of undisturbed beach was designed to permit some erosion at the downdrift coast. A 200 m length of groyne was placed in the middle of the beach. The groyne was functionally designed to trap all sediments coming from the east direction and it was treated as a non-erodible structure. The height of the groyne was 8 m and high enough to prevent sand from over passes the body of the groyne structure. Elsewhere in the model, a 10 m layer of erodible sand was employed. An equidistant grid size of 10 m was used to cover the whole domain area of 1000 x 4000 m. **Figure 3-1** shows the bathymetry and cross-shore bed profile that was used in all simulations performed in this chapter.

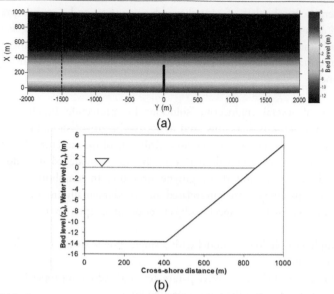

Figure 3-1. Model bathymetry setting. (a) Plan-view of model bathymetry. Dotted line indicates a cross-section of offshore bed profile as presented in panel. (b) Bed level refers to (m+MSL).

3.2.2 Model scenarios and parameter settings

Several model scenarios were developed. Table 1 lists the model scenarios that were setup to run for 50 and 365 morphological days representing conditions before and after the initiation of bypassing, respectively. Models were employed to run for both stationary and non-stationary waves. It should be noted that the XBeach model computed the instantaneous wave energy, E as a function of root mean square wave height, H_{rms}. Therefore, most of the results presented in this chapter refer to H_{rms}. Discussions on the influence of wave angles, wave heights, and sediment grain sizes are provided in **Section 3.3**, **Section 3.4**, and **Section 3.5,** respectively.

Table 3-1: Model scenarios

Case scenarios	D_{50} (μm)	H_{rms} (m)	θ (*deg.*)
Scenario I	500	1.0	20:30:40:50:60:70
Scenario II	500	0.5:1.0:1.5	30
Scenario III	250:500:1000	1.0	30

θ : wave angle; H_{rms} : root mean square wave height;
D_{50} : sediment grain size.

Analytical solutions

An analytical model was used to compare the shoreline evolution and sand bypassing rate with the outputs of the Xbeach model. For the analytical solution, the shoreline changes approach of Pelnard-Considere (1956) was applied. Detailed formulations of this approach were given in **Chapter 2.** The longshore sand transport amplitude was calculated based on the Coastal Engineering Research Centre (CERC) relation (SPM, 1984). The breaking wave angle (H_b) was determined by the iteration of breaking depths, where the breaker index was assumed to be 0.7. Other parameter settings are presented in **Table 3-2.**

Table 3-2. Parameter setting for analytical solutions

Input parameters	Units	Values
Wave height, H_{rms} / H_{sig}	m	1.0/1.4142
Wave period, T_p	s	8.0
Offshore wave angle, θ_o	$deg.$	20: 30: 40: 50: 60: 70
Breaker index, γ	-	0.70
Empirical coefficient, κ	-	0.77
Groyne length, L_g	m	200

$H_{sig} = \sqrt{2}.H_{rms}$

Numerical solutions

The XBeach model was used in this study. The model boundary conditions were given as water level boundary at the seaside and water level gradient (Neumann) boundaries (Roelvink and Walstra, 2004) at the lateral boundaries. The advantage of using the XBeach model was that the model grid for flow and wave could be identical and no disturbances were created at lateral flow boundaries (Roelvink et al., 2009). For sediment transport computation, the Van Rijn formula (van Rijn, 2007) was used with equilibrium sediment grain sizes of 250 μm, 500 μm, and 1000 μm. An avalanching module was activated to account for slumping of sandy material near the waterline. This could allow erosion behind the groyne structure. A higher morphological acceleration factor (morfac) value was applied to reduce the computational effort. This was tested with smaller morfac values (150,100, 50, 20, and 5). The shoreline planform outputs resembled each other (see **section 3.7**). It was assumed that bypassing of sand only began when the groyne tip was fully covered by updrift sand. The numerical model parameters were set almost the same as the parameter in the analytical model, as presented in **Table 3-3.** Further details of the XBeach model description can be obtained in Roelvink et al., (2009; 2010).

Table 3-3. Model parameters for numerical solution

Input parameters	Units	Values
Wave conditions		
wave height, H_{rms}	m	0.5: **1.0** : 1.5
wave period, T_p	sec	**8.0**
offshore wave direction, θ_o	deg.	**200: 210: 220: 230: 240: 250: 260**
Sediment transport		
sediment size, D_{50}	m	0.000250: **0.000500**: 0.001000
onshore transport, *facua*	-	0.2
Morphology		
Morfac factor	-	300

dry slope factor, *dryslp*	-	0.1
wet slope factor, *wetslp*	-	0.1

* The bold items indicate cases where results were compared with analytical models.

3.2.3 Stationary wave of XBeach

The selection of stationary wave was favourable for a simple case problem as such in this study. This may save computational time. Meanwhile realistic results would be expected. Although many test cases have been examined using the stationary wave of XBeach (e.g McCall et al., 2010, Dissanayake, 2012), it is important for this study to ensure that the incoming wave from the offshore to the nearshore zone was reasonably propagated.

The analytical model estimation applied a linear wave model based on a linear wave theory. The linear wave theory included important processes such as shoaling, breaking, wave set-up and wave set-down. This theory (known as Airy theory) assumed that wave amplitude was small in relation to water depth and wave length. It was assumed that the longshore component was uniform, thus restricting the model application to only shore normal wave. In XBeach, the stationary wave was calculated row by row in an iterative way. The interval between the wave modules was called through the *waveint* parameter. This option may substantially reduce computational times.

Figure 3-2 shows the comparison of stationary wave height between the analytical linear wave model and the numerical XBeach model. A perfect agreement of the wave heights between XBeach and linear wave models could be seen from the deepest depth to a shallower depth near the breaking zone. In the linear model the wave height to water depth ratio is constant at 0.7, whereas in XBeach the wave breaking parameter gamma of 0.7 leads to more gradual wave dissipation and on average a lower wave height to water depth ratio.

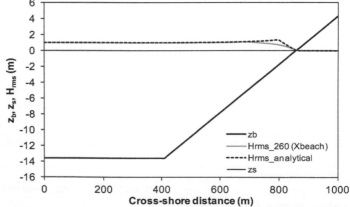

Figure 3-2. Stationary wave comparison between XBeach and analytical linear wave models

3.3 Results and discussions: Effect of wave angles

In general, a wave propagates from deep water to a near-shore area at a certain angles. The high and low angles of a wave lead to large and small sediment transport along the shore, respectively. The wave angle was defined as the angle between the wave crest and the prevailing wave direction. Extensive research was carried out to investigate the relationship between wave angles and longshore transport rates. For instance, Aston and Murray (2006) provided an excellent discussion on this relationship using various well-known sediment transport formulations. Their study found that the maximum undisturbed longshore transport rate was achieved when the wave angles are in the range of 30° to 50°.

In this study, wave angles ranged from 10° to 70°, representing the smallest and largest angles. All cases used a constant wave height of 1 m amplitude, a period of 8 s, and a uniform grain size (D_{50}) of 500 μm. Both stationary and non-stationary waves were considered. Comparisons of shoreline evolution and bypassing volume between the XBeach model and analytical Pelnard model were implemented.

3.3.1 Shoreline evolution - Before initiation of bypassing

A comparison between the shoreline pattern from the numerical XBeach model and the analytical model is presented in **Figure 3-3**. The resulting shoreline pattern for the analytical model was determined using **Equation 2.11 and 2.13** in **Chapter 2**. The equation re-reads as:

$$y(x,t) = \alpha_b \left(\frac{2\sqrt{at}}{\sqrt{\pi}} e^{\left(\frac{x}{2\sqrt{at}}\right)^2} - x.erfc\left(\frac{x}{2\sqrt{at}} \right) \right) \qquad t < t_{fill} \qquad (3.10)$$

$$y(x,t) = L_g .erfc\left(\frac{x}{2\sqrt{at}_{bypass}} \right) \qquad t > t_{fill} \qquad (3.11)$$

For fair comparison, the closure depth used in the analytical model was determined based on the numerical model prediction. In the numerical model, the computation of closure depth was done by subtracting the final bed profile from the initial bed profile. The first small difference ($\Delta_{bed} \approx 0$ m level) from the offshore boundary indicated the limit of morphological change. This limit indicated the closure depth. The models were successfully run for 50 morphological days with constant wave angles and no other forces involved. In this case, there was no assumption of bypassing.

Shoreline evolution on the updrift side of structure showed a good agreement between the analytical Pelnard model and the XBeach numerical model both for stationary wave and Jonswap spectrum boundary conditions (see **Figure 3-3**). The stationary wave of the XBeach models - generated smooth shoreline pattern due to its regular sea state. Despite the fact that the Jonswap wave spectrum produced slight variations, it corresponded well with the analytical model of

Pelnard. The shoreline predicted by analytical models was poorly represented on the downdrift beach. Consequently, the numerical XBeach model simulated the realistic shoreline pattern with a small rate of erosion near the groyne. This shoreline pattern was comparable to the analytical downdrift shoreline.

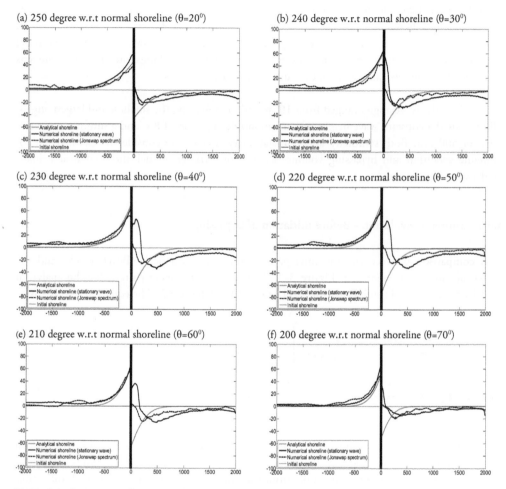

Figure 3-3. Comparison of numerical shoreline (XBeach) and analytical shoreline (Pelnard) for two different wave boundary conditions for 50 morphological days simulation. The x-axis and y-axis indicate alongshore distance (m) and offshore distance (m), respectively.

Unlike the Jonswap wave spectrum boundary condition, the stationary wave condition was not realistic. The stationary wave created a unique feature immediately behind the groyne, particularly when approaching wave angles of 40°, 50° and 60° (see **Figure 3-3-c, d, e**). This feature formed a like sand wedge and started to grow progressively in line with the advancement of the accretion adjacent to the updrift barrier. Sorenson (1978) concluded that the formation of sand wedge was a result of wave propagation from the opposite direction. However, this simulation was purely based on single wave direction only.

The formation of this unique feature occurred, as a result of wave diffraction and the role of water circulation near the shadow region. In order to understand what was really happening in the leeside of the groyne, the horizontal current pattern for both the stationary wave boundary condition and Jonswap wave spectrum boundary condition were plotted (see **Figure 3-4**).

For the stationary wave, the reverse current on the leeside of the groyne was responsible for the previously observed formation of the sand wedge (see **Figure 3-3**). Evans (1943) used a hypothetical model with a small headland to explain the behaviour of this current. Some of the material stirred up by this current was deposited on the bed just to leeward. In addition, Pattiaratchi et al., (in press) discussed the pattern of eddy circulation on the leeside of the groyne and found that eddy was strongest when the incident wave angle was 45°. On the other hand, the small magnitude of current of Jonswap wave mimics a realistic shoreline pattern downdrift of the beach.

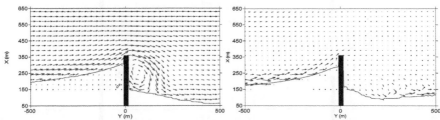

Figure 3-4. Horizontal current pattern at t = 50 morphological days for the incident wave angle of 50°. Left panel represents the stationary wave boundary condition and right panel corresponds to the Jonswap wave spectrum boundary condition. The initial shoreline was located 150 m from the x-origin.

3.3.2 Shoreline evolution - After bypassing

The simulation's duration was expanded to one year in order to allow the bypassing of sand over the groyne tip. Within this timeframe, the morphological sand bypassing occurred in the downdrift zone. A significant amount of sand had been transported to the downdrift beach. Six model simulations were carried out to represent small (30°), medium (50°), and large (70°) incident wave angles. For these scenarios, both the stationary wave boundary and the Jonswap wave spectrum boundary were used.

Figure 3-5 shows the contour plot of the resulting bathymetry after 50 days and 365 days of numerical simulations both for stationary wave and Jonswap wave spectrum. Compared to the 50-day simulation, the updrift shoreline in the 365-day simulation has migrated offshore resulting in wider beach area. The Jonswap model produced more shoreline advancement and beach recession updrift and downdrift of the coastline than the stationary wave. The erosion rate on the downdrift zone that was initially maximized due to water circulation motion (see **Figure 3-5**) has been reduced significantly as the updrift sand started to bypass and filled the eroded area. Consequently, a sand wedge grew immediately next to the groyne.

The underwater bed contour line migrated seaward moving parallel to itself while maintaining its shape. This migration coincided with the analytical solution assumption. An obvious difference between these wave boundary conditions was the erosion pattern on the downdrift zone. The Jonswap model constituted significant erosion prior to bypassing while the stationary model showed a small magnitude of shoreline recession (see **Figure 3-5**, left panels). However, the trend changed after the bypassing takes place by showing some similarities in bed contour changes both for stationary and non-stationary waves.

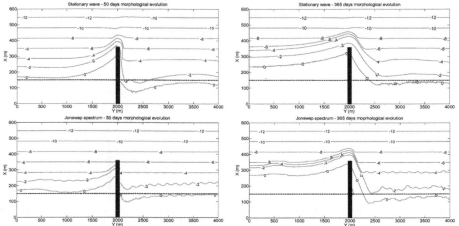

Figure 3-5. Numerical shoreline evolution for stationary wave and Jonswap spectrum wave boundary condition after 50 and 365 morphological days, respectively. A dotted line indicates the initial shoreline.

3.3.3 Initiation of sand bypassing process

The mechanism of groyne bypassing is shown in **Figure 3-6**. It applies the Jonswap wave spectrum boundary condition for a wave angle of 30°. Initially, the alongshore transport caused the sand to accumulate updrift of the groyne (**Figure 3-6a**). Then the sand moved around the groyne tip (**Figure 3-6b**) and subsequently merged with the downdrift beach (**Figure 3-6c**), causing slight accretion in the previously eroded area. This bypassing process resembled the process of sand bypassing over the headland tip as discussed by Evan (1943) and Short and Masselink (1999).

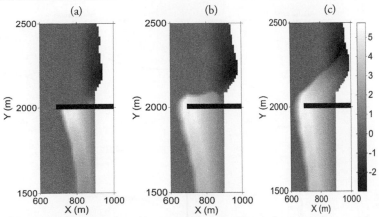

Figure 3-6. The mechanism of groyne sand bypassing. The colour bar on the right indicates erosion (dark) and deposition (light) in meters.

Although their explanations referred to a small-scale headland, the modelled process behaved similarly to that of the groyne structure. Indeed, the bypassing process was found to influence the stability of embayed beaches (Silveira et al., 2010). A comparison of the shorelines generated by the Pelnard' analytical solution and the output of the numerical XBeach model for a one year simulation is graphically presented in **Figure 3-7**. Only the Jonswap wave model was used to generate these outputs since the results of stationary wave models over-predict the analytical model.

In contrast to the result of the 50-day simulation, the average shoreline of the numerical simulation did not correspond with the analytical solution of Pelnard, in the 365-day simulation. Updrift of the groyne, the shoreline simulated by the numerical model was only in a good agreement with the Pelnard model for the locations within 500 m of the groyne. The models showed large differences in areas further away from the groyne. However, the case with a larger wave angle (wave angle 200°) showed otherwise (see **Figure 3-7**, lower panel). Since the alongshore scale was limited, wave-driven currents force the sediment to fill the updrift coast and thus prograde the coastline. As the shoreline approached the groyne tip, large volumes of sand were transported to the downdrift coast. This continuous bypassing process seemed to occur if the groyne tip was at water depths shallower than closure depth (within surf zone), which agrees with the findings of Silveira et al., (2000) as well as Short and Masselink (1999).

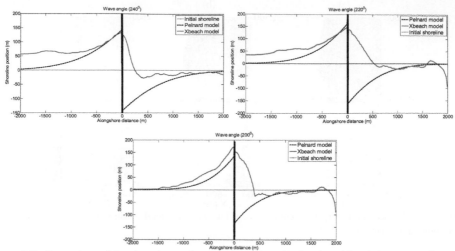

Figure 3-7. Comparison of shoreline changes between the analytical Pelnard model and the numerical XBeach (Jonswap spectrum wave) for three different wave angles.

The Pelnard model predicted a more landward shoreline position in the updrift zone and a greater magnitude of erosion at the downdrift shoreline. Similar to other process based models (for example, Delft3D), the XBeach model demonstrated significant sand bypassing. High volumes of sand flowed around the groyne and settled just downstream of it. This process cannot be taken into account by analytical models. In these models, the groyne blocks all sediment. Consequently, the computed shoreline planform on the updrift side of the groyne yields an almost identical prediction to the Pelnard model (Roelvink & Walstra, 2004).

The factor that could influence the deviation between models was the selection of closure depth. Ravens and Sitanggang (2006) revealed that the percentage error of the GENESIS coastline model output relative to the analytical model in term of shoreline palnform was somewhat smaller when closer to the groyne and greater when further from the groynes. This finding showed that, one of the key factors in analytical solutions was the estimation of closure depth.

Hallermeier (1981) provided an estimation of the closure depth, which was a function of wave climate. This equation limits the usefulness of using closure depth as a morphodynamic boundary and reduces its applicability to computing the closure depth (Capobianco et al., 1997). In two-dimensional situations, the Hallermeier closure depth provided robust estimates of the limit of closure for individual erosion events for time scales up to one year (Nicholls et al., 1998). However, the estimation of closure depth only considered a cross-shore redistribution of sediment, which made it invalid in areas that were rapidly accreting due to longshore supply of sand (Nicholls and Birkemeier, 1997).

Selecting the appropriate closure depth was complex and made it difficult to compute the shoreline change due to a large intervention. The advance in the shoreline was larger for smaller closure depth. This was because the shoreline's advance was calculated by distributing the locally incoming sediment along the whole active profile (Capobianco et al., 2002). If closure depth was underestimated, a large change in the shoreline would be predicted (Capobianco et al., 2002). However, the influence of closure depth on shoreline evolution could be understood by testing various shapes of seabed profiles and discovering how these profiles would influence migration at that depth.

3.3.4 Sand bypassing volume

In order to determine the volume of sand bypassing in front of the groyne in the numerical simulation, the following relationship was applied:

$$\int_0^t \int_{x_{initial}}^{x_{end}} S_y dx dt = \sum_{i_t=1}^{nt} \left[\sum_{i_x=1}^{nx} S_{y(ix)} \Delta x \right] \Delta t \qquad (3.12)$$

where
S_y = total sediment transport (m²/s);
Δx = grid spacing (m); and
Δt = time step (s).

The total sediment transport rate in front of the groyne (S_y) was integrated over the offshore distance (x) at any time. **Table 3-4** shows the cumulative sediment transport for each wave angle within one year of morphological time together with analytical bypassing calculated by the Pelnard method. In all cases, the bypassing volume predicted by the numerical model was much lower than the analytical solution of Pelnard. One of the possible reasons was a difference in the sediment transport formulation; the Pelnard model used the CERC formula (SPM, 1984), whereas the XBeach model applied the Van Rijn sediment transport formula (Van Rijn, 2007). Another possible explanation of the large differences was that the CERC formulation is known to over-predict the computation of longshore transport (e.g. Smith et al., 2009). The CERC formula can be used successfully if parameter K is properly calibrated, see **Section 3.8.1**.

A model such as the Khampuis model (Kamphuis, 1991) which includes wave period, a factor that influences breaker type could give better comparison with the XBeach model. Furthermore, the Khampuis formula gives transport rate as a function of H_b^2, whereas transport rate using CERC formula is a function of $H_b^{5/2}$. Therefore, the CERC formula could simply give results higher than the Khampuis model. The computed bypassing volume using the Khampuis model for the wave angle of 30^0, 50^0, 70^0 was 405,000m³, 587,000m³, and 295,000m³, respectively. Although the analytical bypassing volume computed using Kamphuis method is lower than the CERC method, yet the bypassing volume is greater than the XBeach model. However, other parameters may also lead to the large differences between XBeach and Pelnard model i.e closure depth and filling time. These factors are further discussed in **Section 3.8**.

Table 3-4. Bypassing volume of the XBeach and Pelnard models

Wave angle approaches (°)	Pelnard model (m³)	XBeach model (m³)
1. Small; 30°	860,000	156,000
2. Medium; 50°	1,000,000	198,000
3. Large; 70°	516,000	171,000

Bypass volume of the XBeach model is calculated for non-stationary waves.

The numerical bypass volume was maximized at a medium wave angle, which follows the analytical trend of sand bypassing and longshore sediment transport (not presented here). Ashton and Murray (2006) detailed the usability of various types of sediment transport formulations. They revealed that longshore sediment transport amplitude was maximized for angles between 35° and 50°, depending on the formula. This finding showed that bypassing volume coincides with the magnitude of the longshore sediment transport, assuming that the beach material was not carried away by the cross-shore transport process and that the wave propagated in one prevailing direction.

Figure 3-8 shows the temporal distribution of the bypassing volume over a one-year simulation. The XBeach simulations, using the Jonswap spectrum wave boundary for all representative wave angles, predicted a slower transition of bypassing time, which was around 100 to 150 days. The time-dependent bypass volume rose up to 400 m³ to 450 m³ depending on the direction of the wave. In contrast, the analytical models estimated that sand bypassing occurred for a much longer time period than the numerical models. In order to capture the full recovery of longshore transport and to reach an undisturbed area, the model domain should be extended further upstream and much further downstream. In addition, the model time step should also be kept longer in order to be able to observe the equilibrium stage of bypassing.

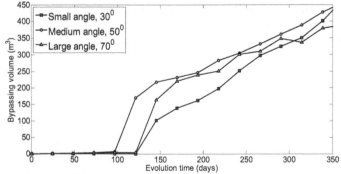

Figure 3-8. XBeach numerical bypass volume during a one-year simulation using different wave angles.

Shoreline planform and sand bypassing volume between the XBeach and Pelnard models were compared. The application of the XBeach model showed some promising results. The shoreline prediction before the initiation of bypassing coincided with the analytical Pelnard model. The circulation of wave-driven currents due to diffraction process was responsible for the formation of the sand wedge, thus prograding the coastline. The analysis of bypass volume showed that the XBeach model predicts a lower amount than the Pelnard model, exhibiting large deviation of shoreline planform for the three representative wave angles.

3.4 Results and discussions: Effect of wave heights

The influence of wave heights was assessed. Wave heights ranged from 0.5 to 1.5 m, representing the low, moderate, and high wave energy conditions that commonly observed in near shore areas. In all cases, the incoming wave angle was held constant at 30° with a period of 8 s. A uniform grain size (D_{50}) of 500 μm was used throughout the model. Other parameters settings were similar to the model setup described in **Section 3.2.2**.

The results presented herein are not meant for comparison with the analytical models. However, the discussion mainly focuses on comparing the effects of different parameters used within the same model. The extent to which the variation of these parameters influences the shoreline pattern and sand bypassing process are further elaborated below.

3.4.1 Shoreline pattern

Figure 3-9 shows the numerical shorelines for three different wave heights at the end of day 50. In all cases, wave heights contributed to a significant change in the shoreline pattern. This could be seen both on the updrift and downdrift sides of the groyne. Increasing wave heights led to shoreline advancement on the updrift side of the groyne and intense erosion on the downdrift side of the groyne. Diffraction caused a sheltered area on the leeside of the groyne to form. Low waves (H_{rms} = 0.5 m) tend to prograde the updrift shoreline, thus slowing the erosion process at the downdrift beach. This phenomenon occurred differently for high waves (H_{rms} = 1.5 m). For H_{rms}=1.5 m, sand was accreted on the leeside of the groyne. The accretion was caused by the initiation of sand bypassing. High waves caused more turbulence in front of the groyne, resulting in faster sand bypassing than the low waves.

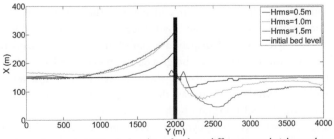

Figure 3-9. Comparison of numerical XBeach shorelines for three different wave heights at the end of day 50.

Extending the simulation time to 365 days initiated sand bypassing. **Figure 3-10** shows shoreline positions for three different wave heights at the end of one morphological year. Similar to the case before bypassing, the shoreline planforms varied significantly according to wave height.

The shoreline advancements just behind the groyne structure for the moderate and high wave heights, i.e. H_{rms}=1.0 m and 1.5 m, proved that bypassing of sand occurred. In contrast to the case prior to bypassing, the sheltered area on the leeside of the groyne disappeared by day 365. The sheltered area was filled by the sand, which has bypassed the groyne tip. However, the low wave (H_{rms}= 0.5 m) showed shoreline pattern, which contrasts the moderate and high waves just behind the groyne. Since there was no bypassing, the shoreline planform retreated.

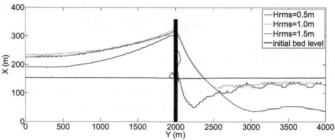

Figure 3-10. A comparison of numerical XBeach shorelines for three different wave heights. The plots show conditions at the last time step (365 days).

3.4.2 Sand bypassing process

Figure 3-11 presents the morphological evolution of sand bypassing around the groyne tip for three different wave heights: low wave (H_{rms}=0.5 m), moderate wave (H_{rms}=1.0 m), and high wave (H_{rms}=1.5 m). At the beginning of day 10, there was no bypassing process for low and moderate waves. Nevertheless, sand was already trapped on the updrift side of the groyne. There was less trapped sand for the low wave than the moderate wave. However for the high wave, numerical results showed that sand had accumulated in front of the groyne tip without any bypassing occurring.

After day 50, the low wave had caused significant sand accretion on the updrift side of the groyne. On the other hand, the moderate and high waves showed that sand had bypassed the groyne tip. For the moderate wave, the sand had merely attached to the downdrift side of the groyne. However, the high wave had dispersed a significant volume of sand to the downdrift beach. By the day 100 with the low wave, the groyne was blocking a large volume of sand. This resulted in significant sand accretion, similar to what occurred with the low wave height. Yet, the results of low wave showed sand did not bypass the groyne tip. On the contrary, when simulating using both the moderate (H_{rms}=1.0 m) and high (H_{rms}=1.5 m) wave heights, results showed that most of the sand already bypassed the groyne tip. As a consequence, sand was transferred to the downdrift side of the groyne to fill the erosional spot.

At day 300, sand still had not bypassed the groyne tip at low wave height. However, sand accreting on the updrift structure was noteworthy. Moderate and high waves already caused sand to bypass the groyne. On the downdrift side of the groyne, shoreline erosion was intensified at all wave heights. Beaches behind the groyne were receding remarkably before the start of bypassing.

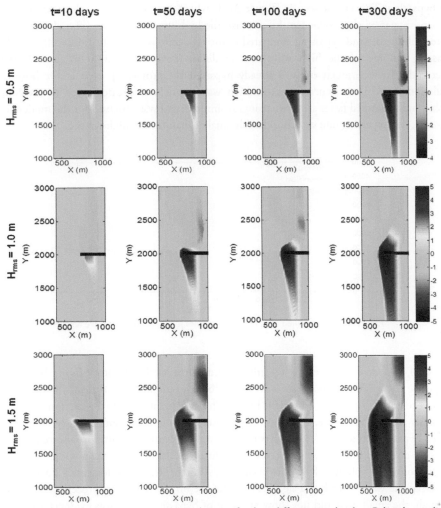

Figure 3-11. The process of sand bypassing around groyne for three different wave heights. Colour bars indicate erosion (blue) and deposition (red).

3.4.3 Sand bypassing rate

Figure 3-12 shows the computed bypass volume over one morphological year for three different wave conditions: H_{rms} = 0.5, 1.0, and 1.5 m. It should be noted that the bypassing volume was determined based on the cumulative sediment that crosses over the groyne tip. The bypassing volume of XBeach was determined using **Equation 3.1** as shown in **Section 3.3.4.**

As shown in **Figure 3-12**, the computed volume of bypassed sediment increased with the amplitude of wave height. The bypassing volume of low wave height i.e. H_{rms} = 0.5 m showed zero bypassing. This can be seen in **Figure 3-11**, where even at 300 days, there was no evidence that sand bypassed the groyne tip. In contrast, the high wave amplitude (H_{rms}=1.5 m) showed a greater volume of sand bypassing compared to the moderate wave height (H_{rms}=1.0 m). The total bypassing volume of the high wave and moderate wave was 45,000 m³ and 18,000 m³, respectively. In fact, much of sand already bypassed the groyne tip even at the beginning of simulation time. This occurred because high waves generated greater speed in the longshore current and promoted faster sand bypassing. A similar trend was observed for the moderate wave. However, the bypassing rate was considerably smaller than for the high wave.

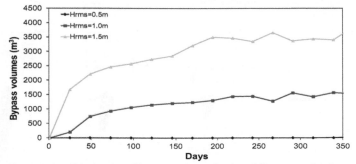

Figure 3-12. Calculated sand bypasses volumes for three different wave heights.

Overall, the wave height exerts a significant influence on the amount of sand bypassing over the groyne tip. High and moderate waves tend to transport more sand from the updrift beach, around the groyne tip to the downdrift beach. This was because updrift accretion occured faster for higher waves. Therefore, the coastline orientation migrated more quickly shoreward and allowed bypassing to sooner. The high wave caused more sediment transport also resulting higher bypassing rates. Additionally, high waves broke further offshore, creating a wider surf zone. For these reasons, the volume of sand bypassing was much great than for small waves. This also shows that the wave force had a significant influence on the bypassing process, as stated earlier. Although, the small wave did not show any sign of bypassing, there was evidence that sand was continuously trapped on the updrift side of the groyne.

3.5 Results and discussions: Effect of sediment grain sizes

Various sediment grain sizes (D_{50}) were employed. Two extra calculations were performed with fine and coarse sediment sizes. The sediment grain sizes were ranged from 250 μm to 1000 μm representing fine, medium and coarse sand. In all cases, the wave angle was constant at 30° with a wave height of 1.0 m and a period of 8 s. Other parameters settings were similar to the model setup described in **Section 3.2.2**. It should be noted that shoreline plots fluctuated slightly because of numerical instability. Nevertheless, the shoreline patterns in the model behaved as predicted.

3.5.1 Shoreline pattern

Figure 3-13 shows a snapshot of the shorelines at the end of day 50, which were computed using the XBeach model for three different sediment grain sizes: fine (D_{50}=250 μm), medium (D_{50}=500 μm), and coarse (D_{50}=1000 μm). The results did not differ as much as with the change in wave heights. On the updrift side, fine sediment caused a greater seaward movement of the shoreline than the medium and coarse sediments. This indicated that fine sediments create a wider beach area. Therefore, if the beach was not wide and sand bypassing did not occur, the downdrift beach would suffer from intense erosion.

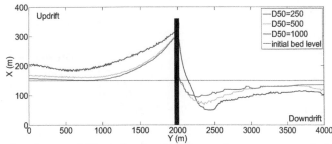

Figure 3-13. Snapshot of numerical XBeach shoreline for different sediment grain sizes at the last time step i.e. 50 days.

By day 50, accretion of fine sediments on the updrift side of the groyne occurred faster than in the cases of medium and coarse sediments. This sped up the bypassing process. This raised the bed level adjacent to the groyne. Raising of the bed occurs whenever sand accretes adjacent to the groyne structure (see D_{50} = 250 μm). However, the medium and coarse sediments did not show much difference of shoreline patterns neither on the updrift nor the downdrift side of the groyne.

The morphological evolution of seabed changed after one year is illustrated in **Figure 3-14**. At the end of the day 365, the fine sediment showed obvious seaward movement of the shoreline and greater sand accretion on the leeside of the groyne.

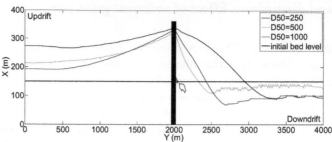

Figure 3-14. Snapshot view of numerical XBeach shoreline for different sediment grain sizes at the last time step (365 days).

The shoreline position in front of the groyne tip shifted slightly offshore. This indicated that much of sand was deposited and remained in front of the groyne tip (see **Figure 3-15**). The accumulation of sand in front of the groyne proved a seaward migration of shoreline. In the fine sediment scenario, the beach width in the area far from the updrift side of the groyne was two times wider than the original beach line. Generally, the beach width of fine sediment was larger than the medium and coarse sediments. Although the volume of accreted sand behind the groyne for the medium and coarse sediments was not as large as the fine sediment, the results showed that sand has bypassed the groyne tip.

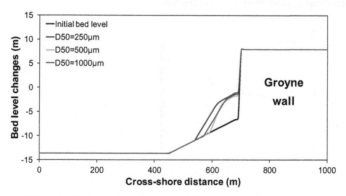

Figure 3-15. Accreted sand in front of the groyne head over 365 days.

3.5.2 Sand bypassing process

While low wave heights decelerate the process of bypassing, fine sediments seems to increase the speed of bypassing around the groyne tip. **Figure 3-16** presents the evolution of sand bypassing for three different sediment grain sizes along 300 morphological days.

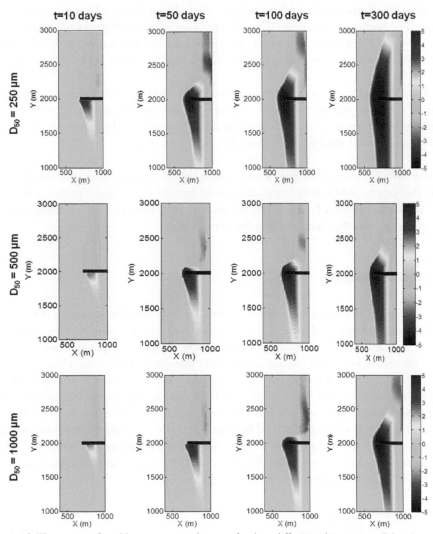

Figure 3-16. The process of sand bypassing around groyne for three different sediment sizes. Colour bars indicate erosion (blue) and deposition (red).

By day 10, some sand had been deposited on the updrift side of the groyne regardless of sediment grain size. Fine sediment accumulated more sand on the updrift side of the groyne and sand had already reached the groyne tip. For the moderate and coarse sediments, there was less sand trapped on the updrift side of the groyne.

After day 50, the coarse sediment (D_{50}=1000 μm) showed no evidence of sand bypassing around the groyne tip. Instead, sand progressively accreted on the updrift side of the groyne. On the other hand, fine sediment (D_{50}=250 μm) showed a considerable amount of sand bypassing the groyne tip. Consequently, a large volume of sand was deposited on the leeside of the structure. The medium size sediment (D_{50}=500 µm) showed that sand just begun moving around the groyne tip. At this point, the process of bypassing for the fine and medium size sediments resembles the bypassing process of high and medium waves (see **Figure 3-15** for comparison).

By day 100, sand bypassed the groyne tip for all sediment sizes. Fine sediment caused greater deposition downdrift side of the groyne compared to the medium and coarse sediments. The numerical result of coarse sediment showed that sand only bypassed the groyne tip and deposited a small volume of sand attached on the downdrift side of the groyne.

By day 300, fine sediment showed a considerable volume of sand accumulated on the updrift side of the groyne. Additionally, sand dispersions beyond the groyne tip were visible. This process was similar to that of medium and coarse sediments. However, these medium and coarse sediments had a lower bypassing rate than the fine sediments.

3.5.3 Sand bypassing rate

The temporal pattern of the sand-bypassing rate for three different grain sizes: fine sediment (D_{50}=250 µm), medium sediment (D_{50}=500 µm), and coarse sediment (D_{50}=100 µm) is presented in **Figure 3-17**. It clearly shows that the computed bypassing volume had two different patterns of bypassing. For fine and medium sediments, sand began to bypass the groyne tip immediately, while coarse sediment had a short delay before the sand bypassing commences. The bypassing volume of the fine sediment was far higher than the two others sediment classes. The total bypassing volume over one year course of simulations for the fine, medium and coarse sediments was 41,000 m³, 18,000 m³ and 8000 m³ respectively. While the bypassing volumes of medium and coarse sediments showed a constant trend, the bypassing volume of fine sediment exhibited erratic behaviour especially after day 200. This might have been due to the large amount of sand that was deposited in front of the groyne tip.

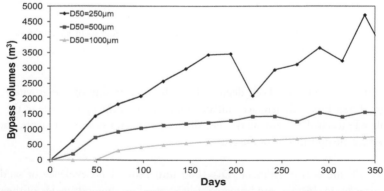

Figure 3-17. Temporal pattern of XBeach bypassing rate for different sediment grain sizes.

3.6 Effect of upstream and downstream model boundaries

The effect of upstream model boundary and downstream model boundary is investigated to verify that the accumulation of sand updrift of the groyne structure is not affected by the upstream model boundary. In **Figure 3-11** and **Figure 3.16,** we showed the time evolution plots of deposition and erosion for different wave heights and sediment grain sizes, respectively. In those figures, the shoreline advancement updrift of the groyne due to sand infilling was observed as a result of a limited alongshore distance at the upstream model boundary. If the alongshore distance is extended further upstream, this phenomenon could be prevented to occur. In order to prove this hypothesis, we extended the alongshore distance of the model to 8 km *i.e* 4 km on each side of the model boundary. Model was run with a H_{rms} wave height of 1.0 m, wave period of 8 s, and wave angle of 30^0.

Figure 3.18 shows the plots of sedimentation and erosion for an extended alongshore model domain at the end of 300 morphological days.

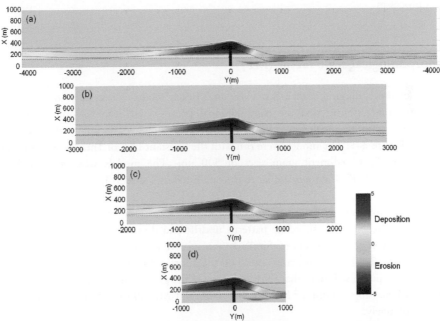

Figure 3-18. Erosion/sedimentation patterns for an extended alongshore model distance after 300 days of simulation. In all panels iso-contours (3 m intervals) are contoured in the background. Black dotted lines are the initial bed countour lines.

In **Figure 3.18 (d)**, it is obvious that a huge accumulation of sand is only observed at the groyne from a distance of 1 km from the groyne structure. As the coastline moves seaward and reaching to the end of the groyne's tip, a wider beach develops. This phenomenon was observed in **Figure 3-11** (see middle panel- rightmost). Away from the groyne structure *i.e.* at a distance of 3 km from the groyne, the results from the extended alongshore model showed that the beach appeared to remain stable and undisturbed (see **Figure 3.18(b)**). The coastline is hardly changed as the upstream model boundary did not have a significant influence on the model result. Just rightly at the upstream boundary there was a positive effect which is resulting to a small deposition of sand which is believed due to the wave shadowing effect. In addition to that effect, the coastline slightly moved seaward from its original position.

Immediately behind the groyne structure, as sand began to bypass the groyne, it filled the erosional spot leeside of the groyne. The beach appeared to slightly receding but remains unchanged further down to the downstream area. The extended erosion area on the downdrift side is obviously seen in the model as shown in the full alongshore model scale of **Figure 3.18 (a)**. This can be explained by the difference in wave setup and advection terms that lead to a slow build-up of the longshore current and these factors was described by Roelvink and Reniers (2012). Another possible explanation is the profile response in the downdrift area. The longshore transport can be established if the eroded profiles react by shifting horizontally (Roelvink and Reniers (2012). In XBeach, this depends both on avalanching and the cross-shore transport processes.

3.7 Morphological scale factor (MF)

The use of morphological scale facotr (MF) in XBeach model is necessary to speed-up the computational time, noting that the computational process of Xbeach is quite heavy, thus requires longer computational time. Further this study employes a schematised model as illustrated in **Figure 3-1**. The models were run with the constant H_{rms} of 1m, Tp 8 s, wave angle of 30^0 , and uniform sediment grain sizes of 500 μm. The models were run for 50-days morphological period. This setup is similar to the one listed in **Table 3-3**.

Figure 3-19 shows the results of shoreline evolutions for three different morfac values. In all cases, model results show the shoreline patterns updrift and downdrift sides of the groyne are coincided, regardless of morfac values. Similarly, cumulative erosion and deposition patterns as shown in **Figure 3-20** show insensitiveness of the morfac value to the morphological bed changes. The computed sediment that initially bypassed the groyne shows that in all simulations, the predicted bypassing volume of MF 20, MF 100, and MF 300 is 7184 m^3, 6248 m^3, and 9856 m^3, respectively.

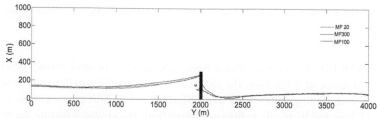

Figure 3-19. 50-morphological days shoreline evolution with different MF values.

Ranasinghe et. al (2011) developed a method for a priori determination of the highest morfac value that is suitable for a given simulation. **Equation 3.13** describes a criterion for the priori determination of morfac criteria i.e.,

$$CFL_{MF} = \frac{C_{bed}MFdt}{dx} < 1 \tag{3.13}$$

By inserting C_{bed} as ~0.001 m/s, MF (300), dt (0.301 sec), dx (10 m), the CFL_{MF} is about 0.0093, still far lower than 1. The CFL_{MF} is dependent on the hydrodynamic time step (dt). The hydrodynamic time step in X**B**each is automatically computed based on the Courant criterion and the value is considerably low in order to guarantee a stable computational process. The use of high morfac in this particular case study is reasonable since all the forcing are relatively mild and constant throughout the simulation.

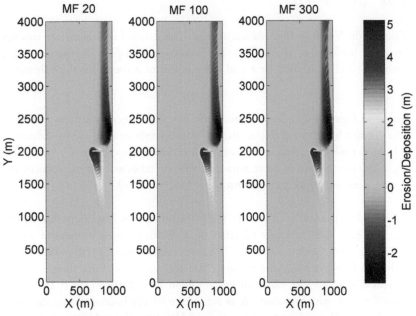

Figure 3-20. Cumulative erosion and deposition with different MF values.

3.8 Further investigation of sand bypassing parameters

Comparing the results between the analytical model and numerical model, specifically in term of bypassing rates brought forth some discrepancies. Therefore, it was worth investigating some of the important parameters that may have contributed to sand bypassing processes and caused deviations between the two modelling approaches. These discussions mainly concerned computing of the longshore transport rates; the influence of closure depth on the initiation of bypassing time; and differences between stationary and non-stationary waves of XBeach.

3.8.1 Longshore sand transport rate

Longshore transport amplitude (Q) was the key parameter for calculating sediment bypassing around groyne structures. Although there were many transport formulas, the use of the CERC formulation was still widely applicable and eases computation. Ashton and Murray (2006) provided a review of some existing transport formula (i.e CERC, Khampuis, Bailard, Deigaard) and found that longshore transport was maximized when the approaching wave angle was between 35° and 50° for 2.0 m, 10 s waves depending on the formula. To further substantiate this, a comparison between the CERC formula (SPM, 1984) and the Van Rijn formula (Van Rijn, 2007) was completed. The CERC formula for longshore transport is given in **Equation 3.14**.

$$Q = K \left(\frac{\rho \sqrt{9.81}}{16 \kappa^{0.5} (\rho_s - \rho)(1-n)} \right) H_b^{5/2} \sin(2\alpha_b) \tag{3.14}$$

In XBeach, the Van Rijn formula was applied. The detailed formulas from this approach can be referred to Roelvink et al., (2009). The longshore transport (S_y) was averaged over the simulation time and multiplied by a grid cell size Δx, to account for the volumetric transport rate.

Referring to **Figure 3-21**, the longshore transport (Q) of both relations seems to reasonably follow the classical longshore transport pattern as previously determined by Ashton and Murray (2006). In both models, increasing the wave angles tends to increase the Q. As the breaking wave angle increases, the amount of wave refraction that stretches wave crests reduces wave heights. This, in turn, tends to reduce the Q (Ashton and Murray, 2006).

The calibration parameter K in the CERC formula was important for determining the longshore transport rate. An early selection of K coefficient was introduced with root mean square breaking wave height, i.e. K=0.77 (Komar and Inman (1970)). This value was most commonly seen in many longshore transport rate computations. Since then, many of variations of coefficient K were present, but with median grain sizes. Bailard (1981: 1984) used K in his energy-based model, which presents K as a function of the breaker angle, the ratio of orbital velocity, sediment fall speed, and as a function of H_{rms} at breaking. However, due to the use of a limited data set, the predicted K coefficient in his relationship may be highly variable.

Others have proposed empirical-based relationships (e.g. Bruno, et al., 1980; Dean et al., 1982; Khampuis et al., 1986; Dean 1987) for increasing K with decreasing sediment grain size. However, the empirical relationships resulted from two data sets with K values based on erroneous and questionable field data (Komar, 1988). Komar (1988), after revising the K values of his previous results concluded that K should depend on sediment grain sizes. Del Velle et al., (1993) presented an empirical based relationship for the K coefficients based on median grain sizes, which ranged from 0.4 mm to 1.5 mm). This relationship was also applied with $H_{b\,rms}$. The relationship infers that the K coefficient should decrease with increasing sediment grain size.

Nonetheless, this relationship should be used with care since it was based on limited data and strongly depends on the data from the specific survey area, the Adra River Delta.

Two values of *K* were tested. They were 0.77 (introduced for use with rms breaking wave height ($H_{b\ rms}$) by Komar and Inman (1970) and 0.45 (which depends on the median grain size, D_{50} from Del Velle et al., (1993). The result of longshore transport of *K=0.77* showed that it was greater than the Van Rijn formulation, despite the longshore transport followed the same trend. The latter coefficient gives approximately the same values for longshore sediment transport as the Van Rijn formulation (see **Figure 3-21**). This shows the critical role of sediment grain size in computing the natural longshore sediment transport rate and its effect on computing the sand-bypassing rate.

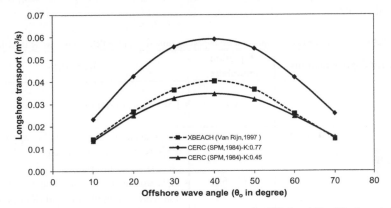

Figure 3-21. Comparison of longshore sediment transport rate using the CERC and Van Rijn formulations.

3.8.2 Closure depth

A direct estimation of the t_{fill} parameter was formulated by Pelnard-Considere (1956) and was useful for the beginning of beach maintenance on the downdrift side of the structure. Since the Larson model did not consider this parameter in estimating the rate of bypassing, the comparison was only between the Pelnard and XBeach models (see **Equation 2.14** of Pelnard and **Equation 2.16** of Larson in **Chapter 2** for comparison). In order to compute the bypassing time in the analytical model, a closure depth (h_c) was required. Since, one of the critical parameters for determining the bypassing of sediment was the time when the sediment started to bypass (t_{fill}), the closure depth was highly important.

Closure depth was defined as the deepest point where the sediment transport was insignificant (Nicholls et al., 1996). Three approaches were used to determine the t_{fill} in analytical model: constant closure depth (h_c= 5m), h_c based on the modified Hallermeier (1981), and h_c directly computed by XBeach model. **Table 3-5** lists the constant h_c and the mathematical equations used for the closure depth computation in the analytical Pelnard model. The t_{fill} was calculated based on the formula given below. By fixing the parameters groyne length (L_g), breaking wave angle (α_b), and longshore transport rate (Q_o), the contribution of h_c to the estimated bypassing time as well as the bypassing volumetric rate could be determined.

$$t_{fill} = \frac{\pi}{4.a} \frac{L_g^2}{\alpha_b^2} \tag{3.14}$$

where shoreline diffusivity $a = \frac{2Q_o}{h_c}$. $\tag{3.15}$

The above relationship for a fixed wave climate revealed that the time required for the shoreline to reach the end of the groyne will increase fourfold if the groyne length is doubled (Larson et al., 1987).

Table 3-5: Closure depth estimation for bypassing time (t_{fill}) in the Pelnard bypassing model.

Closure depth equations	Equation numbers	Remarks
1. constant h_c h_c= 5 m.	(-)	• For all t_{fill}, constant value of h_c is employed.
2. Modified Hallermeier $h_c = 2.28H_b - \dfrac{68.5H_b^2}{gT_p^2}$	(3.16)	• Original closure-depth equation applies an offshore significant wave height (H_s) that exceeds 12 hrs per year. • H_s is replaced by H_b (breaking wave height) following Khampuis' method (Kamphuis, 2000).
3. XBeach $h_c = \dfrac{\sum\limits_{i}^{nx} Zb_i - Zb_{initial}}{\sum\limits_{i}^{nx} nx_i}$	(3.17)	• First small difference ($\Delta z_b \approx 0$ m) from the offshore boundary indicates the limit of morphological change where the closure depth is obtained.

Table 3-6. Computed bypassing time of the Pelnard and XBeach bypassing models.

Wave angle (θ_o)	Bypassing time, t_{fill} (days)			
	Analytical Pelnard model			Numerical models
	h_c=5m	h_c=computed using Hallermeier (1981)	h_c=computed using XBeach	XBeach model
10	1238	755	-	-
20	348	208	-	-
30	184	106	295	121
40	140	76	-	-
50	131	66	204	97
60	162	72	-	-
70	270	98	432	121

(-) *value indicates no simulation runs.*

Table 3-6 shows the computed bypassing time of the Pelnard and XBeach models. The closure depth that was calculated by the XBeach model was used to observe the influence of this parameter in the analytical Pelnard bypassing time (see column 4). In column 5, the bypassing time was determined by plotting of the XBeach bypassing rate against time.

Considering the bypassing time calculated for the Pelnard model (see columns 2, 3, and 4), the closure depth seemed to significantly influence the estimation of t_{fill}. The closure depth essentially restricted the analytical model's computation of bypassing volume as well as the evolution of shoreline. The XBeach model predicted a higher closure depth and bypassing time compared to the modified Hallermeier and constant h_c approaches. The variation of closure depth implied that larger closure depths lead to extensive bypassing times. If closure depth was overestimated, the t_{fill} will be longer and the bypassing volume will be over-estimated and vice versa.

The selection of a "reliable h_c" was essential for the accurate estimation of the bypassing rate. The optimal solution for h_c could not predict the actual closure depth when using simple wave based models such as Hallermeier (1981). However, they could predict distributional properties such as the limits of the closure depth (Nicholls et al., 1998). Additionally, the Hallermeier equation only considered cross-shore redistribution of sediment and excludes the effects of near-shore beach profile translation (Nicholls and Birkemeier 1997). Therefore, in areas that were accreting rapidly due to longshore supply of sand, this equation was particularly ill suited for the analysis.

Additional factors such as randomness in forcing, profile bed slope, and sediment properties were required to apply the h_c concept more widely (Capobianco et al., 2002). An extension of the existing model was required to improve the "weak" capability of the closure depth estimation. Correlating the statistics of depth variation and the statistics of the offshore wave height by using simple approaches may have been advantageous (Capobianco et al., 2002). Based on the XBeach model output, the predicted t_{fill} was much lower than the analytical Pelnard model. The XBeach model also predicted that h_c was substantially larger, leading to longer t_{fill} than in the analytical models. However, the numerical XBeach model showed a reduced influence of the initial bypassing time on the h_c when compared to the analytical model.

3.8.3 Differences between XBeach stationary wave and non-stationary wave solvers

In XBeach, the stationary wave used a forward marching technique, where the equations are solved row by row in an iterative way. The wave module was then set to every wave interval(s) rather than at each time step. This greatly reduced computational time (Roelvink et al., 2010).

On the other hand, the Jonswap spectrum wave was created by a user-input spectrum coefficient. The coefficient was based on the analytical 2D Jonswap-type spectrum and was used to generate the alongshore, varying time series of wave energy and the bound long wave. This option contributed to the realistic generation of second-order bound, directionally spread seas

(Roelvink et al., 2010). To generate the wave group, a procedure to convert parametric spectra to wave energy and a bound long wave was based on a method proposed by van Dongeren et al., (2003).

Two different wave boundary conditions of XBeach were employed: the stationary wave and Jonswap spectrum wave. Three cases were selected to represent small (30°), medium (50°), and large wave angles (70°), which eventually produced six separate models runs. The difference in the temporal distribution of the sand-bypassing rate between the stationary wave and Jonswap spectrum wave models can be clearly seen in **Figure 3-22**.

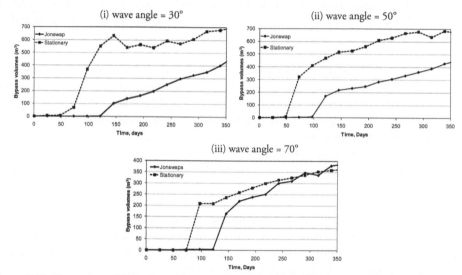

Figure 3-22. Comparison of XBeach sand bypassing curves for three different wave angles. *Jonswaps* refers to non-stationary wave (H_{mo}=1.0 m) and *Stationary* indicates stationary wave.

The stationary wave showed larger bypassing volumes compared to the Jonswap wave spectrum for all three cases. A technical reason behind the bypassing difference results is that the wave energy dissipation due to wave breaking formulations. In stationary wave computation of XBeach, wave energy dissipation is calculated based on Baldock (1998) while in in-stationary wave computation, the formulation of Roelvink (1993) was applied.

However, both models showed similar bypassing mechanisms at the start of bypassing. Both models indicated that sediments started to bypass when there was sand accretion updrift of the structure. This phenomenon resembled the assumption made in the Pelnard model. However the bypassing of the Jonswap spectrum wave began later than the stationary wave. The Jonswap wave boundary condition showed lower bypass volumes than the stationary wave due to a lesser current in front of the groyne structure and along the beach. The behaviour of this current was presented in the **Figure 3-4**. The reduced current was due to the irregular sea state and varying wave energy in the non-stationary wave. The high amounts of sediment bypassing with stationary waves caused a reversed current pattern. Reversed currents are responsible for the formation of the sand wedge on the leeside of the structure as reported by Ab Razak et al., (2013). However, the wedge disappears as the sand starts to bypass the groyne structure. In some areas where the swash cycles and infra-gravity motions play an important role, the application of the non-stationary wave was more realistic than the stationary wave (Roelvink et al., 2009).

3.9 Conclusions

This investigation examined the natural, wave-driven process of sand bypassing across a groyne tip. It covers the effects of wave parameters, i.e. wave heights, wave angles and varying sediment grain sizes, on this process. A comparison between the numerical model and analytical models was completed to observe variations between the results from both models. Additionally, several important bypassing parameters, which may influence the process of bypassing were analysed and discussed. Based on the results presented in this chapter, several detailed conclusions can be drawn.

1. Wave height has an especially large influence on the amount of longshore transport. When waves break further offshore, they cause more transport. Wave breaking causes turbulence and alongshore transport. Turbulence leads to initiation of motion and it prevents settling of suspended sediment. A strong longshore current generated by the high waves can also prevent the settlement of sediment. Also, the surf zone is wider in the case of higher waves. Therefore, the groyne blocks less sediment, increasing the bypass. The bypass also increases rapidly when there is greater turbulence due to higher waves.
2. Smaller grain size allows more sediment transport. Smaller sediment grains also stay suspended longer. Therefore, smaller sediment grains are more likely to bypass the groyne. Smaller grain size resembles the case of higher wave height, in which the width of the surf zone is greater. Thus, this leads to shoreline advancement on the updrift side of the groyne. Alternatively, larger grain size results in slow migration of shoreline and less bypassing over the groyne head.
3. Wave angles seem to influence the shoreline pattern and bypassing rates. There is good agreement between the shoreline planforms of the XBeach model and the analytical Pelnard model. Bypassing volume at a medium wave angle, i.e. 50°, has a maximum rate similar to that of the undisturbed longshore sediment transport.
4. Prior to bypassing, shoreline patterns of the analytical model and the numerical XBeach model are similar, especially when the non-stationary waves are applied. Surprisingly, though the analytical model was developed assuming small wave angles, the agreement remained good even for very large wave angles. There is a clear difference between the analytical model and the numerical model on the downdrift side of the groyne. The XBeach model replicates the downdrift erosion realistically, whereas the analytical Pelnard model produces an unreliable erosion pattern that mirrors the pattern of accretion on the updrift side of the groyne.
5. In the analytical solution, an ambiguity of closure depth parameter seems to significantly influence the sediment-bypassing rate. A selection of higher closure depth leads to higher

bypassing volume and vice versa. A proper definition of closure depth and selection of appropriate approaches can reduce the difference between numerical and analytical solutions.

6. In cases with different wave angles, the stationary wave of the Xbeach model displayed unrealistic behaviour behind the groyne structure. The spit formation behind the groyne maybe only due to high wave angles. Before the initiation of bypassing, a sand wedge is created adjacent to the downdrift side of the structure due to current circulation pattern. However, this only happens at certain incoming wave angles, namely $\theta = 50°$ and $60°$.

7. In case of a large net transport of sediments caused by larger wave height, smaller sediment grains, or larger wave angle the coastline will show extensive accretion on the updrift side and erosion on the downdrift side. In the case of a small net transport, the erosion and accretion will be less.

8. The mechanism of sand bypassing was visualised through the two-dimensional process based XBeach model. In all different cases, a similar bypassing mechanism was achieved. The process begins when the groyne blocks sand at the updrift side. Once the updrift side of the groyne is fully filled, sand then moves around the groyne tip. Afterwards, sand begins to attach to the backside of the groyne before slowly merging with the downdrift beach. The bypassing of sand seems to occur on a sub-aqueous layer as was discussed by Short and Masselink (1999).

References

Ab Razak, M.S. Dastgheib, A. and Roelvink, D. (2013). Sand bypassing and shoreline evolution near coastal structure comparing analytical solution and XBeach numerical modelling, *Journal of Coastal Research*, 65(2): 2083-2088.

Ashton, A.D., and Murray, A.B. (2006). High-angle wave instability and emergent shoreline shapes: 2. Wave climate analysis and comparisons to nature, *Journal of Geophysical Research*, 111: 1-17.

Bailard, J. A. (1981). An energetic total load sediment transport models for a plane sloping beach. *Journal of Geophysical Research*, Vol 86, No.C11, 10938-10954

Bailard, J.A. (1984). A simplified model for longshore sediment transport. *Proceeding of the 19th International Coastal Engineering Conference*, American Society of Civil Engineers, New York, pp 1454-1470.

Bakker, W. T., and Edelman, T. (1965). The coastline of river deltas. *Proceeding of the 9th Coastal Engineering Conference*, American Society of Civil Engineers, New York, pp 199-218.

Bakker, W. T. (1969). The dynamics of a coast with a groyne system, *Proceeding of the 11th Coastal Engineering Conference*, American Society of Civil Engineers, New York, pp 492-517.

Bakker, W. T., Klein-Breteler, E. H. J., and Roos, A. (1971). The dynamics of a coast with a groyne system, *Proceeding of the 12th Coastal Engineering Conference*, American Society of Civil Engineers, New York, pp 1001-1020.

Baldock, T.E., Holmes, P., Bunker, S. and Van Weert, P., (1998). Cross-shore hydrodynamics within an unsaturated surf zone. *Coastal Engineering* 34: 173- 196.

Bruno, R.O., Dean, R.G., and Gable, C.G. (1980). Littoral transport evaluation at a detached breakwater. *Proceeding of the 17th Coastal Engineering Conference*, American Society of Civil Engineers, New York, pp 1453-1475.

Capobianco, M., Larson, M., Nicholls, R.J. and Kraus, N.C. (1997). Depth of closure: a contribution to the reconciliation of theory, practice and evidence. *Proc. Coastal Dynamis'97*, ASCE, pp.506-515

Capobianco, M., Hanson, H., Larson, M., Steetzel,H., Stive, M.F.J., Chateluse,Y., Aarninkhof, S., and Karambas,T. (2002). Nourishment design and evaluation: applicability of model concepts, *Coastal Engineering*, 47: 113-35.

Daly, C.J., Bryan, K.R., Roelvink, J.A., Klein, A.H.F., Hebbeln, D., and Winter, C., (2011). Morphodynamics of embayed beaches: the effect of wave conditions. *Journal of Coastal Research* SI 64, 1003–1007.

Dean, R.G., Berek, E.P., Gable, C.G, and Seymour, R.J (1982). Longshore transport determined by an efficient trap. *Proceeding of the 18th International Coastal Engineering Conference*, American Society of Civil Engineers, New York, pp 954-968.

Dean, R.G. (1987). Measuring longshore transport with traps. *Nearshore sediment transport*, Richard J. Symour, ed., Plenum Press, New York.

Del Velle, R. Medina, R. and Losada, M.A. (1993). Dependence of Coefficient K on Grain Size, Technical Note No.3062, *Journal of Waterway, Port, Coastal and Ocean Engineering*, 119(5): 568-574.

Dissanayake, P.K. (2012) Modelling morphological response of large tidal inlet systems to sea level rise. PhD Thesis, UNESCO-IHE Institute for Water Education, Delft, the Netherlands, 175p.

Evans, O. F. (1943). The relation of the action of waves and currents on headlands to the control of shore erosion by groynes. *Academy of Science for 1943*: 9-13.

Hallermeier, R.J. (1981). A profile zonation for seasonal sand beaches from wave climates. *Coastal Engineering* (4): 253-277.

Jackson, D.W.T., and Cooper, J.A.G. (2010). Application of the equilibrium planform concept to natural beaches in Northern Ireland. *Coastal Engineering*, 57: 112-123.

Kamphuis, J.W. (1991). Alongshore sediment transport rate. *Journal of Waterways, Port, Coastal and Ocean Engineering*, 117(6):624-641.

Kamphuis, J.E., Davies, H.M., Narin, R.B., and Sayao, O.J. (1986). Calculation of littoral sand transport rate. *Coastal Engineering* ,10(1): 1-21.

Kamphuis, J.M. (2000). Introduction to Coastal Engineering and Management, Singapore: World Scientific Publishing, 437 pp.

Komar, P.D. and Inman, D.L. (1970). Longshore sand transport on beaches, *Journal of Geophysical Research*, 75(30): 5914-5927.

Komar, P.D. (1988) Environmental controls on littoral sand transport. *Proceeding of the 21th International Coastal Engineering Conference*, American Society of Civil Engineers, New York, pp 1238-1252.

Klein, A.H.F., Ferreira, Ó., Dias, J.M.A., Tessler, M.G., Silveira, L.F., Benedet, L., de Menezes, J.T., and de Abreu, J.G.N. (2010). Morphodynamics of structurally controlled headland-bay beaches in southeastern Brazil: A review. *Coastal Engineering*, 57: 98-111.

Larson, M., Hanson, H., and Kraus, N.C. (1987). Analytical solutions of the one-line model of shoreline change. *Tech. Rep. CERC-87-15,USAE-WES*, (Vicksburg, Miss.: Coast. Eng. Rest. Clr.).

Larson, M., Hanson, H., and Kraus, N.C. (1997). Analytical solutions of one-line model for shoreline change near coastal structures. *Journal of Waterway, Port, Coastal, and Ocean Engineering*, 123 (4): 180-91.

LeMehaute, B., and Brebner, A. (1961). An introduction to coastal morphology and littoral processes. (Rep. No. 14), Civil Eng. Dept., Queens Univ. at Kingston, ant., Canada.

McCall, R.T. de Vries, J. T. Plant, N.G van Dongeren, A.R., Roelvink, J.A Thompson, D.M and Reniers, A.J.H.M. (2010). Two-dimensional time dependent hurricane overwash and erosion modeling at Santa Rosa Island, *Coastal Engineering*, 57(7): 668-683.

Nicholls, R.J. Birkemeier, W.A., and Hallermeier, R.J. (1996). Application of the depth of closure concept. *Coastal Engineering 1(25)*: 1-14.

Nicholls, R.J., and Birkemeier, W.A. (1997). Morphological and sediment budget controls on depth of closure at Duck, NC. *Proc. Coastal Dynamics 97*, ASCE, pp. 496-505.

Nicholls, R.J., Birkemeier, W.A., and Lee, G., (1998). Evaluation of depth of closure using data from Duck, NC, USA. *Marine Geology* ,148():179-202.

Pattiaratchi, C., Olsson, D., Hetzel, Y., and Lowe, R. (*in press*). Wave-driven circulation patterns in the lee of groynes. *Continental Shelf Research* (2009), DOI:10.1016/j.csr.2009.04.011.

Pelnard-Considere, R. (1956). Essaidetheoriedel evolution desforms derivagesen plage desableetdegalets. *Fourth Journeldel' Hydralique, lesenergiesdela Mer, QuestionIII,*, Rapport No.1 289–98.

Raabe, A.L.A., Klein, A.H.F., González, M., and Medina, R. (2010). MEPBAY & SMC: Software tools to support different operational level of headland-bay beach in coastal engineering projects. *Journal of Coastal Research*, 57:213-226.

Ranasinghe, R., Swinkels, C., Luijendijk, A., Roelvink, D., Bosboom, J., Stive, M., and Walstra, D. (2011). Morphodynamic upscaling with the MORFAC approach: Dependencies and sensitivities. *Coastal Engineering,* 58 (2011): 806-811.

Ravern, T.M., and Sitanggang, K.I. (2006). Numerical modeling and analysis of shoreline change on Galveston Island. *Journal of Coastal Research,* 22 (0): 000-00.

Roelvink, J.A. (1993). Surf beat and its effect on cross-shore profiles. PhD Thesis, Technical University of Delft, the Netherlands, 150 p.

Roelvink, J.A., and Reniers, A.J.H.M. (2012). A guide to modelling coastal morphology. Advances in Coastal and Ocean Engineering Vol.12. World Scientific, Singapore. 274p.

Roelvink, D. and Walstra, D.J. (2004). Keeping it simple by using complex models. *The 6th International Conference on Hydroscience and Engineering* Brisbane, Australia, pp 1-11.

Roelvink, D., Reniers, A.J.H.M., van Dongeren, A., van Thiel de Vries, J., McCall., R., and Lescinski, J. (2009). Modelling storm impacts on beaches, dunes and barrier islands. *Coastal Engineering,* 56 (11-12): 1133-52.

Roelvink, D., Reniers, A.J.H.M., van Dongeren, A., Vries, J.T., Lescinski, J., and McCall, R. (2010). XBeach model description and manual. UNESCO-IHE Institute for Water Education.

Short, A. D. and Masselink, G. (1999). Embayed and structurally controlled beaches, In: Short, A.D.(ed)., *Handbooks of Beach and Shoreface Hydrodynamics.* Chicester: John Wiley & Sons. pp. 230-249

Smith, E.R., Wang, P., Ebersole, B.A., and Zhang, J. (2009) Dependence of total longshore sediment transport rates on incident wave parameters and breaker type. *Journal of Coastal Research,* 25(3): 675-683.

SPM (1984). 2 vols.(4th Ed.; Washington, D.C.: USAE-WES Coast. Engrg. Res. Ctr., U.S. Govt. Printing Ofc.)

Silveira, L.F., Klein, A.H.F. and Tessler, M.G. (2010). Headland bay beach planform stability of Santa Catarina State and of the Northern Coast of Sao Paulo State. *Brazilian Journal of Oceanography,* 58(2): 101-122.

Sorenson R. (1978). *Basic Coastal Engineering.* USA : John Wiley & Sons, Inc. 324p.

van Dongeren, A., Reniers, A., Battjes, J., and Svendsen, I. (2003). Numerical modelling of infragravity wave response during DELILAH, *Journal of Geophysical Research Oceans,* 108(9): 1-19.

van Rijn, L. C. (2007). Unified View of Sediment Transport by Currents and Waves. I: Initiation of Motion, Bed Roughness, and Bed-load Transport. II: Suspended transport. III: Graded beds. *Journal of Hydraulic Engineering,* 133 (6): 649-775.

Weesakul S., Rasmeemasmuang T., Tasaduak S., and Thaicharoen C. (2010) Numerical modelling of crenulated bay shapes. *Coastal Engineering* 57:184-193.

CHAPTER 4

Headland structural impact on the morphodynamics of embayed beaches

This chapter presents the morphodynamic investigation of embayed beaches through the impact of structural headlands. An XBeach model was applied to predict the surf zone current circulation pattern and to predict morphological features of three different embayment models for low-moderate-high wave energy events. The mechanism of transporting sediment outside the surf zone was proven through the implementation of virtual drifters on different scale of embayment basins. The formation of central rip currents in embayed beaches is linked to the presence of a sand bar, while topographical headland rips developed adjacent to the headland boundary are caused by the geological control of the headland structure itself. The effect of moderate and high waves has resulted in a decreased number of central rips in a longer embayment thus limiting the beach circulation to the cellular type. Whether wave breaking occurs outside or inside the embayment determines the initiation of large scale rip currents, megarips. Nevertheless, the characteristics of surf zone current circulation in all cases presented in this study complies with the description of the theoretical embayment scaling parameter (δ'). The irregular wavegroup forcing was incorporated in the embayed beach models and model results showed its presence contributes to the rapid morphological evolution of embayed beach bathymetry. The wave directional spreading was found to affect the morphological bed changes of embayed beaches. Surf zone rip currents, specifically headland rips were found responsible to transport the floating materials outside the embayment basin.

Parts of this chapter were published in:

(i) Ab Razak, M.S., Dastgheib, A., Suryadi, F.X., Roelvink, D. (2014). Headland structural impacts on the surf zone current circulations, *Journal of Coastal Research*, SI 70: 65-71.

(ii) Ab Razak, M.S., Dastgheib, A., and Roelvink, D. (2014). Morphodynamic investigation of embayed beaches through the impact of structural headlands, Abstract in : *NCK Days 2014*. Delft, the Netherlands

4.1 Introduction

Embayed beaches comprise 50 % of the world's beach. Commonly bounded by two natural headlands or shore connected breakwaters, these curvilinear-shaped bays are exposed to wave refraction/diffraction processes that may create a sheltering zone at the leeside of the headland, depending on the direction of the incoming waves. A complex pattern of current circulations leads to a dynamic behaviour of embayed beach morphology. This may be one of the reasons for the scarce research development for embayed beach systems. An obvious characteristic of embayed beaches compared to open straight beaches is its surf zone circulation. Headlands and engineering structures like groynes will impact the beach and surf zone through their influence on wave refraction and attenuation, thus limiting the development of longshore currents, rips, and rip feeder currents (Short and Masselink, 1999).

Surf zone rip currents (e.g topographically controlled headland rips, normal beach rips and megarips) in embayed beach systems are responsible for the mechanism of cross-shore sediment exchange (Loureiro et al., 2012; Short, 1985; Coutts-Smith, 2004). Rip currents are generated as a result of alongshore variations in wave height of the incoming waves (Bowen, 1969). The generation of rip currents in embayed beaches is linked to the embayment size and variation of wave heights. Short and Masselink (1999) were the first to investigate the structural impact on surf zone current circulation. The novelty of embayment scaling parameter (δ) that was firstly developed by Short and Masselink (1999) and later was improved by Castelle and Coco (2012) is applicable to characterize the current circulation in different embayment scales under low wave energy conditions.

During storms, high wave energy may modify current patterns in embayed beaches. Beach rips which are normally developed in the middle of an embayment may increase in space to initially develop strong seaward rip currents. The variation of rip current strength was found to be directly related to incident wave height variations over periods of a few minutes. Strong rip currents developed in the middle embayment are associated with higher waves and waves of longer period (Mckenzie, 1958; Huntley et al., 1988). However, Shepard and Inman (1950) opposed this opinion in which the minimum rip current strength is associated with high waves. This is not absolutely clear, but a possible explanation may be associated with the fact that Shepard and Inman (1950) were observing large scale rip currents on an open coast while observations of Mckenzie (1958) and Huntley et al., (1988) were made in relatively enclosed bays. The generation of a central rip current may be through wave refraction around the bay. Long crested waves will refract to approach the shoreline in the bay at an oblique angle everywhere, except at the centre of the bay. This obliquity will drive feeder currents, inside the surf zone, towards the centre of the bay, where the convergence might be expected to create seaward-flowing rip currents. Wright et al., (1978) explained the formation of central rip currents in a highly compartmented beach driven by storm waves. The large rips exhibit a very low frequency and are very destructive in their effects. One or two rips may extend up to several kilometers seaward in highly compartmented beaches, like the one observed in Palm beach (Cowell, 1975) and Cronulla Beach (Lees, 1977).

Despite the fact that the number of studies on morphodynamics in embayed beaches have been increasing recently (e.g Gallop et al., 2011; Ojeda et al., 2011; Loureiro et al., 2012), less attention is given to embayed beach morphodynamic modelling, with the exception of Reniers et al., (2004), who found a relationship between rip current spacing and directional spreading. Furthermore, studies on surf zone morphodynamics in embayed beaches are restricted to a limited number of numerical studies, specifically for low wave energy conditions only. While surf zone retention is well documented on open beaches (e.g. Spydell et al., 2007; Reniers et al., 2009, it has never been addressed in detail on embayed beaches (Castelle and Coco, 2013).

In this study a 2DH process based XBeach model (Roelvink et al., 2009; 2010) was applied as a numerical tool to assess the confidence of the model in examining relevant morphodynamic processes that exist in embayed beach systems. Although the XBeach model is principally designed to predict dune erosion due to storm impact (e.g Roelvink et al., 2010; Callaghan et al., 2013; Splinter et al., 2013), the model can also be applied for small scale coastal engineering problems (e.g Ab Razak et al., 2013).

The main objective of this study is to understand the morphodynamic behaviour of rip channel systems at different embayment scales based on the embayment scaling parameter (δ) for a range of wave heights. A research gap between this study and other researchers (e.g Castelle and Coco 2012) is the inclusion of higher wave energy events and wave group-scale forcing on the morphodynamic modelling of embayed beach systems. Comparison results were made between the model with a stationary wave and the model with an instationary (wave-group forcing) wave. Additionally, the mechanism of cross-shore sand bypassing was investigated through the implementation of virtual drifters that were initially placed in the nearshore zone area.

4.1.1 Embayment scaling parameter (δ)

The embayment scaling parameter (δ) was used as a base guideline to predict the degree of headland impact on the surf zone current circulation. This parameter relates the embayment configuration to the incident breaking wave conditions according to **Equation 4.1** (Short and Masselink, 1999):

$$\delta = \frac{S^2}{100 \, C_l H_b} \tag{4.1}$$

where S is the embayment shoreline length , C_l is the embayment width and H_b is the breaking wave height. Based on δ, three main beach circulations have been identified. On beaches with no headlands or obstacle structures, normal surf zone circulation prevails ($\delta > 19$). When the embayment size and shape begins to increasingly influence surf zone circulation, by causing longshore currents to turn and flow seaward against each headland, while still maintaining some normal beach circulations away from the headlands, a transitional circulation exists ($8 < \delta > 19$). If the headlands are closer together or if wave height increases, the entire beach circulation may be impacted by the headlands. At this stage, a topographically controlled large rip current (Short, 1985), prevails ($\delta < 8$). This type of current generally promotes the cross-shore exchange of sediment transport. (e.g Loureiro et al., 2012; Short, 1985; Coutts-Smith, 2004).

However, the parameter proposed by Short and Masselink (1999) has several limitations as has been addressed by Castelle and Coco (2012). They assumed that wave energy is redistributed along the whole wet-dry contour of the embayment. If the headland length is greater than the surf zone width, the headland impact will be overestimated. Also, the amount of energy dissipated against the headland in most cases is small compared to the one dissipated along the beach. For that reasons, Castelle and Coco (2012) have established a non-dimensional embayment scaling parameter (δ') which considers the surf zone width (Xs) that fits into embayment length (L) as presented in **Equation 4.2**

$$\delta' = \frac{L \, \gamma\beta}{H_s} \tag{4.2}$$

where γ is the breaking parameter, β is the nearshore slope and H_s is significant wave height. The degree of headland impact and circulation type in embayed beaches can still be estimated following Short and Masselink (1999) but with an improved description on the number of developing rips between normal and transitional beach circulations. If the number of rips that exist in the middle of the embayment is greater than four (nrip> 4) the beach is considered normal, while if nrip< 4 transitional beach circulation prevails.

On the other hand, the cellular beach circulation is said to take place when there is a presence of headland rips at both sides of the headland and only one or two rips in the middle of the embayment. This parameter indeed requires further improvements in terms of beach curvature, prevailing wave angle, geometry of the headlands and directional wave spreading. Nevertheless, the degree of embayment predicted by δ' has proved to be consistent with several observations of embayed beaches (Castelle and Coco, 2012). Their embayment scaling parameter was used in this study to quantify the number of resulting rip currents and rip channels on different alongshore scale of embayment basins.

4.1.2 Variability of embayed beach conditions

Knowledge on embayed beach conditions are essential to comprehend our initial understanding on the future morphodynamic modeling of embayed beaches. Embayed beaches can be characterised by three main beach conditions *i.e* (i) dissipative beach, (ii) intermediate beach and (iii) reflective beach, (Wright and Short, 1984) just like the unconstrained sandy beaches. Breaking wave height (H_b), wave period (T) and grain size (as defined by sediment fall velocity, W_s) are those three parameters that characterised the three main beach conditions in a wave dominated beaches. Gourlay (1968) used the combination of these parameters to form the dimensionless fall velocity (Ω) and is calculated as:

$$\Omega = \frac{H_b}{W_s T} \qquad\qquad (4.3)$$

Wright and Short (1984) adapted Ω to classify the three main beach conditions. If $\Omega < 1$, beaches tended to be reflective, when $\Omega > 6$ they tended to be dissipative, and in between $\Omega = 2\text{-}5$ they were what they called intermediate.

Dissipative beaches are characterized by wide, low gradient surf zones across which spilling breakers dissipate their energy. Beaches receive realtively high waves (>2.5 m), preferably short wave period andrequire fine sand. Beaches are fronted by concave upward nearshore profiles and wide flat inshore profiles. Topography is much more complex and varied than in the case of reflective beaches. One or more bars, there dimensional inshore topography, and different scale of rip cells are frequently present. **Figure 4-1(a)** show the typical cross-section of a dissipative beach.

Intermediate beaches refer to beaches that are intermediate between high energy dissipative and the lower energy reflective beaches. The most obvious characteristic of intermediate beaches is the presence of a horizontally segregated surf zone with bars and rips. Moderate to high waves (0.5m to 2.5m), fine to medium sand, and longer wave periods are those environmental conditions that characterized intermediate beaches. Intermediate beaches are the most common beach type and occur both as single barred beaches, particularly on open swell coasts, and commonly as the lower energy inner bar or bars of multi barred dissipative beaches, particularly on short period sea coasts. Because of large range of wave heights, substantial range of morphodynamic character may develop on these beaches., For that reason, intermediate beaches are categories into four beach state known as longshore bar and trough, rhythmic bar and beach, transverse bar and rip and finally low tide terrace. The erosion of intermediate beaches is dominated by the presence of the seaward-moving rip currents. Short (1985) introduced four types of rips i.e accretion rips, erosion rips, topographic rips and megarips. **Figure 4-1(b)** show the typical cross-section of an intermediate beach.

Reflective beaches are characterized by the wave energy that is reflected from the beach face. Distinguishing features include: (i) a linear, low gradient nearshore profile composed of fine sand; (ii) a pronounced steep composed of the coarsest available material and (iii) a steep, linear beach face surmounted by a high berm; (iv) well-developed beach cusp and (v) surging breakers with high run up and minimum setup. Features such as ridge and runnel topography, swash bars, and inshore troughs are consistently absent from such reflective systems. Large scale inshore (subaqueous) rhythmic topographies are completely absent. Similarly, inshore circulations cells and normal beach rip currents are rarely present except during storms. However, topographical rips still present due to headland structures. **Figure 4-1(c)** shows the typical cross-shore profile of a reflective beach.

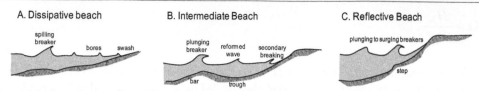

Figure 4-1. Cross-shore profiles of three different types of beaches. (a) Dissipative beaches and (b) intermediate beaches and (c) reflective beaches (From Komar, 1998).

4.1.3 Wave-group effects on embayed beach circulation

In the nearshore zone, a significant part of the wave energy is at frequencies that lower than those of incident swell and wind waves. This motion is referred to infragravity wave motion. Infragravity waves with periods between 20 s to several minutes are generally associated with the groupiness, or the beat, of the incident waves (Munk, 1949; Tucker, 1950). The infragravity wave that is generated by the incident short wave groups can be identified by three types : (1) bound long waves that are forced locally by the wave-groups and travel at the group velocity of the wind waves (2) leaky waves, i.e. reflected infragravity waves at the shoreline that radiate away from the surf zone; travel in the offshore direction, and (3) trapped long waves (edge waves) that cannot escape from the shoreline due to strong refraction.

Researches on wave-group effects are widely documented in several scientific publications and that their presence in the nearshore region influence surf zone current circulation and rip channel dynamics in embayed beaches. Reniers et al., (2004) concluded that wave group-scale forcing influences rip spacing on an embayed beach system. One of the significant findings in their study was that the infragravity wave energy (generated by wave-groups) causes rapid morphological bed changes. The infragravity wave also tends to smoothen the final bathymetry. Likely, the final bathymetry produced by their models shows strong similarities with the model results obtained by Damgaard et al., (2002) despite the morphological evolution speed was significantly faster than the morphological evolution speed obtained in the model of Damgaard et al, (2002) . On the other hand, several researches e.g McKenzie (1958) and Chappel and Wright (1978) found that velocities in the rip channel frequently pulse at low frequencies as a response to infragravity waves. It is evident also that wave-groups can transport suspended materials out of the surf zone (Renier et al., 2009; Renier et al., 2010; MacMahan et al., 2010; Castelle et al., 2010).

4.1.4 Two dimensional (2DH) process based XBeach model

The XBeach model consists of formulations for short wave propagation, shallow water equations, sediment transport and bed update. In this study, both stationary and instationary (wave group) wave solvers were used. In the case when stationary wave is applied, the wave action and roller energy balances are not integrated in time, but are iterated until the equilibria across the model domain are found. In the case when an instationary wave is used, the incident short wave energy is generated for every directional bin, at every offshore grid boundary point, at every time step. Infragravity water fluxes at every offshore boundary grid point at every time steps are generated as well. These detailed processes consume expensive numerical computations, thereby a parallel computing approach is highly needed for models when an instationary wave is used.

The Van Thiel-Van Rijn sediment transport model was used in combination with the advection-diffusion equation to compute total sediment transport, which can be used to update bathymetry. The bathymetry is updated with the use of a (low) morphological scale factor (Morfac) to accelerate the computational time. The avalanching module is activated to account for the slumping of sandy material near the foreshore area.

4.2 Model Setup

4.2.1 Hypothetical embayed beaches and model setup

Two groynes were used to represent headland structures with a length of 300 m while the alongshore beach length was varied (L = 500, 1500 and 4000 m). The groyne length is long enough to prevent any sand bypassing, functionally designed for deeply enclosed beaches. A single barred coast was used as an initial bathymetry with an average slope of 0.01. **Figure 4-2** shows the model embayment geometry and cross-shore bed profile.

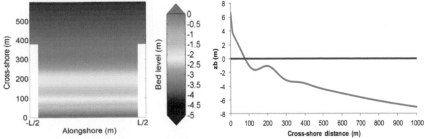

Figure 4-2. Model bathymetry and a nearshore barred bed profile. Red line indicates shoreline limit.

The rip channel is assumed to exist only between the shoreline and the bar line (Gallop et al., 2011). A sand bar located 100 m from the shoreline was placed with an amplitude of 1 m following the approach of Roelvink (1993). Wave height of 1 m, 2 m and 4 m associated with a period of 10 sec were imposed at the offshore seaward boundary. The breaking formulation of Baldock (1999) was applied in all simulations with the breaking coefficient of 0.73, a common value for regular waves. Bed and suspended load sediment were computed based on Vanthiel-Vanrijn formula with a uniform sediment grain size of 100 µm.

Neumann boundary was used as a lateral boundary at both right and left sides of the computational domain. Models were forced to run for ten morphological days with a morphological acceleration factor (Morfac) of 5. The shorter time scale addressed in this study is necessary to describe the dynamic environments of embayed beaches characterized by a variety of processes and a range of complex behaviour resulting in the presence of morphological patterns and the formation of rips (e.g. Holman et al., 2006; Gallop et al., 2011; Ojeda et al., 2011, Castelle and Coco, 2012).

4.2.2 Embayed beach circulations and test cases

Embayed beach circulation model of Castelle and Coco (2012) was used to develop model test cases for this study. **Figure 4-3** shows the beach circulation types and test cases. The δ is calculated based on **Equation 4.2**. A typical current circulation for different ranges of embayment sizes under low wave condition was first evaluated, followed by an investigation on the effect of increasing wave heights. All tests performed in this study are graphically presented in **Figure 4-3.**

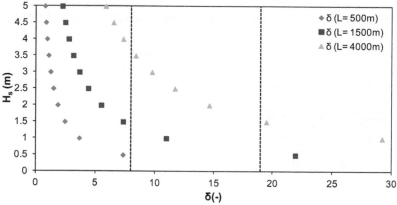

Figure 4-3. Embayment scaling parameter for different wave heights and embayment lengths. Vertical dotted lines delimit different beach circulation states.

4.3 Results and discussions: Effect of stationary wave

4.3.1 Typical rip channel system on embayed beaches : Low wave energy condition

The morphodynamic rip channel of low wave energy was first investigated. For this analysis, a stationary wave of 1 m, period of 10 sec and wave angle perpendicular to the shore $\theta = 0^0$ and 5^0 was applied at the offshore boundary. The alongshore beach length was varied (L = 500, 1500 and 4000 m) and corresponded to embayment scaling parameter of δ= 3.7 (cellular), 10.9 (transitional), and 29.2 (normal), respectively.

Figure 4-4 b,d,f shows the predicted morphodynamic pattern of embayed beaches for an oblique wave of 5^0. These figures clearly show the influence of headland structure as well as embayment length on the development of surf zone rip currents and rip channels. Central rip currents appear in all ranges of embayment. For the L=1500 m, three main rip channels are developed in the middle of the embayment with persistent topographical rips adjacent to each headland boundary. The presence of this central rip current is due to the accumulation of water in the shoreface, so that the water has to flow seaward to escape. Shore parallel wave propagating to the shore breaks over the bar, carrying extra water into the shoreface near the beach (wave setup). The bar and the waves block this accumulation of water from moving seaward over the bar. Instead, a rip current forms where a break (gap) opens up in the bar. Here, water that is piling up flows along between the bar and the beach and then turns and flows seaward through the gap in the bar. As the embayment length increases, the number of central rip channels also increases as shown in **Figure 4-4 (f)** for L=4000 m.

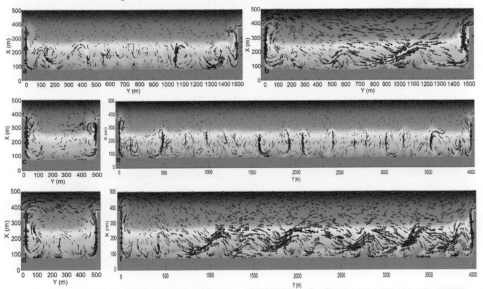

Figure 4-4. Morphodynamics of embayed beaches for different alongshore distances with $H_{rms}=1$ m, $T_p=10$ sec at the last time step. Red bold arrows indicate the incoming waves in two different directions.

This case is similar to the case of an unconstrained beach with no structure in between. The only difference is that headland rips occur at both sides of the structures. The shape of the rip seems to be dependent of the approaching wave angle. On the other hand, for the shorter embayment length (L=500 m), the middle rip current hardly appears although there is evidence of a small developing rip at the centre of the embayment. Still, headland rips do occur.

The morphodynamic surfzone responses of the shore normal waves provide reasonable results, similar to the oblique wave. **Figure 4-4 a,c,e** shows the predicted morphodynamic pattern of three embayed beaches for the shore normal wave case. Rip currents and rip channels are observed in the middle of the larger embayment i.e L= 1500 m and L= 4000 m. The perpendicular flow of the seaward currents provides a clear distinction between the shore normal wave and an oblique wave. In all cases, the development of rip channels complies with the description of theoretical embayed beach circulation.

The morphological development of a rip system can be inspected through the 2D numerical model. This may increase our understanding on the initial formation of a rip channel in embayed beaches. **Figure 4-5** shows the example of the evolution of a rip channel for the case of longer embayment with shore normal waves. During the first two days, the sand bar starts to evolve but merely with small perturbations. A scouring channel develops at both headland extremities due to a strong rip flowing adjacent to the headland (headland rips). After day four, rip channels in the middle of the embayment continuously develop with seven rip currents draining the entire beach. More channels develop after day six, breaching the gap of the bar, carrying more water that is flowing back to the sea. While the rip channels progressively develop, the detached sandbars tend to move shoreward and finally merge with the beach. This happens as the shore normal wave pushes the sand consistently towards the coast. At the end of day 10, the rip channels are still present with four rips existing but their patterns are not as significant as those observed during day four.

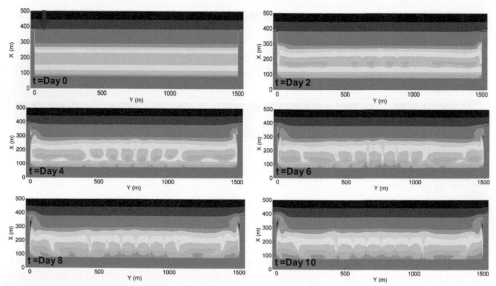

Figure 4-5. Morphological evolution of a rip channel system in a longer embayed beach (L = 1500 m). Red arrows in each panel indicate shore normal wave direction with a H_{rms} of 1m.

This model evolution also shows the transition between the longshore bar-trough state (day 4) and the transverse bar and rips state (day 6) which was observed in the field (see **Figure 4-6**). This was supported by Symonds et al. (1997), who described the slow onshore migration of sand bar to onshore during the period of low waves while developing the longshore variability and rip channels.

Figure 4-6. Rip channels and transverse bars with a rhythmic shoreline under low wave condition. Courtesy of Gallop et.al (2011).

4.3.2 Effect of increasing wave heights: The transition from a transitional to a cellular beach state

We further investigated the morphodynamic behaviour of a rip channel system when the wave height was increased to 2 m. In this case, an embayed beach with a length of 1500 m was only tested as this embayment indicates the transition between the transitional state to a cellular state based on embayment scaling parameter, δ, as shown in **Figure 4-3**.

The case of L=500 m was tested also, but is not presented here as it does replicate the similar behavior pattern of 1 m wave height, with only headland rips occurring at both headland extremities. Additionally, the contribution of increasing wave height for the shorter embayment is not unique as it is already categorized in a cellular beach state as illustrated in **Figure 4-3**. The nearshore bed profile is kept constant similar to the previous analysis.

Figure 4-7 shows the evolution of bed level changes and current circulation pattern for the moderate wave height (H_{rms} = 2 m). Based on δ, it is categorized under cellular beach circulation as δ = 5, less than the boundary limit of cellular beach state. This indicates that the number of central rip currents will be decreasing.

At the beginning of simulation (day 4), there are seven rips developing in the middle of the embayment. These rips persistently remain at the same position until day 8. At the end of the simulation, the number of developing central rips decreases to only five rips thus increasing the mean alongshore rip spacing. The number of central rips for this case (H_{rms}=2 m) is against the finding of low wave condition, i.e H_{rms} = 1 m.

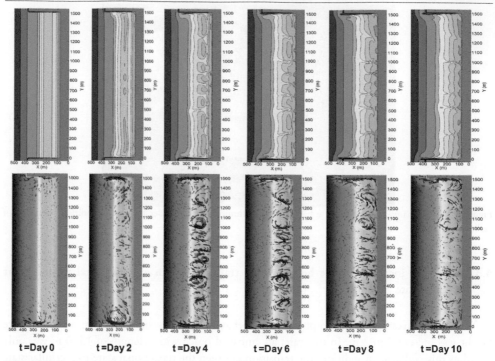

Figure 4-7. Evolution of bed level changes (upper panels) and current circulation pattern (lower panels) of H_{rms} 2 m under shore normal waves. Bed level is represented as colorbar in **Figure 4-2**. In all upper panels iso-contours (0.5m intervals) are contoured in the background.

To further investigate this we recorded the number of central rip currents that are developed during each day. The summary of mean rip spacing both for H_{rms} =1 m and H_{rms} = 2 m is presented in **Table 4-1**. Obviously, increasing wave height leads to a less number of rip currents developing in the middle of the larger embayment. This supports findings of Short (1985), who noted that under increasing waves, the alongshore rip spacing increases (decreasing number of rip currents) with the rips becoming more intense. Rips may configure by shifting, disappearing and more can re-appear (Short, 1985). The number of rip currents increases over the course of simulation inferring a decrease in alongshore mean rip spacing. As the channel starts to develop more channels appear (day 4) and slowly decrease as the bar slowly migrates inshore (day 10).

Table 4-1: Central rip currents developed during each day

Time evolution [days]	0	2	4	6	8	10
	Numbers of central rip currents [-]					
H_{rms} = 1.0 m	-	5	8	9	9	8
H_{rms} = 2.0 m	-	3	7	7	7	5
H_{rms} = 4.0 m	-	5	5	5	5	3

4.3.3 Inclusion of extreme wave condition

The inclusion of extreme wave condition in investigating the morphodynamics of an embayed beach was incorporated. Referring to **Figure 4-3**, the beach circulation is categorized as cellular thus limiting the presence of rips in the middle of the embayment. For this particular case, an additional run for a wave height of 4 m was carried out to represent an extreme wave condition. The bed profile was extended offshore to a greater depth to prevent any disturbance at the seaward boundary.

Figure 4-8 shows the morphodynamic behavior of an embayed beach for wave conditions of H_{rms} 4 m and shore normal wave ($\theta = 0^0$). The current circulation pattern and bed level changes are more or less similar to the case of H_{rms} 2.0 m with only a slight re-positioning of the rip channel. On the other hand, at the alongshore location of 800 m, an intense rip channel has developed, scouring the seabed up to -4 m, thus indicating the presence of a strong rip current in the middle of the embayment. In our simulation, the breaking wave height on the sand bar just before the bed is updated is rather low, merely 0.6 m for the wave height of 4 m, slightly above that for the moderate wave height of 2 m. This implies that the wave initially breaks further offshore, losing its energy and breaking again near the bar with less energy. The evolution of rip current rip channel for this particular extreme wave event shows a migration of the rip channel. This can be explained through the number of developing central rips as presented in **Table 1**. On day four, significant rip channels appear with five rips developed in the middle of the embayment. The number of central rips remains the same for the following days but is reduced to merely three rips when approaching day 10.

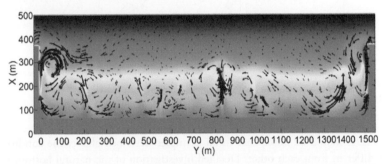

Figure 4-8. Surf zone current circulation and morphological bed changes of an embayed beach with H_{rms} of 4.0 m. Bed level is represented as colorbar in **Figure 4-2.**

There are several studies that reported different surf zone current circulation behaviour in embayed coasts, which were not reproduced in our modelling study. For instance, Short (1985) reported the existence of large central rips in three Sydney beaches due to the influence of nearshore and adjacent embayment topography that prevented the development of a fully dissipative state by inducing wave refraction and persistent longshore gradients in surf zone dynamics. One or two megarips may drain the entire long embayed beach as the wave height increases, especially during extreme wave events (see **Figure 4-9-right panel**).

In a different case, the storm oblique waves promote the formation of a meandering rip feeder current towards the east side of the headland. The rips sustain the same position, not experiencing significant alongshore movement as they remain topographically controlled. Likewise, Silva et al., (2010) showed that the accumulation of water at the centre of the beach leads to the formation of a strong seaward rip. As the headland structures are close to each other and the shore-normal wave energy is evenly distributed along the beach, the rip current develops in the middle of embayment. Once the circulation is established, the increasing wave height would only strengthen and weaken the current velocities without modifying the circulation pattern (Silva et al., 2010). On the other case study, Loureiro et al., (2012) described the formation of one large central rip developing roughly on the centre of the embayment during high energy events (see **Figure 4-9-left panel**) in Amoreira embayment, southwestern coast of Portugal. The beach is approximately 600 m long and generally exposed to energetic north-westerly waves. The formation of megarips is not related to the breaker gradients at this beach, but specifically due to the nearshore topographic control.

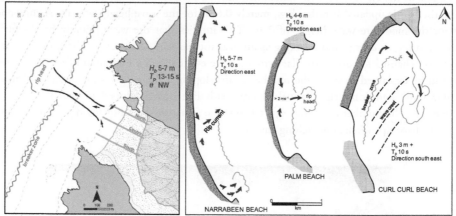

Figure 4-9. Schematic maps of megarip circulation. Left panel presents small Amoreira embayment (Portugal) during storm conditions (Loureiro et al., 2012). Right panel shows large central rip circulations in the middle of three large embayments in Sydney region (Short, 1985).

Nevertheless, it should be noted that in all examples described above, the condition of each study area is different from each other. Detailed investigation of the natural bottom topography at those particular study areas is highly recommended. In fact, small bathymetry irregularities can influence surf zone hydrodynamics, inducing wave gradient driven rip circulations (Calvette et al., 2007; MacMahan et al., 2008; Dalrymple et al., 2011). On the other hand, wave breaking conditions either inside or outside the embayment also determine the generation of megarip currents. Short (1985, 2007, 2010) proposed a breaking wave height of 3 m as a threshold value for the generation of megarip currents. Further analysis on the formation of megarip currents should be investigated to understand the physical processes that govern their generation.

4.3.4 Limitations of the embayment scaling parameter

In unconstrained coasts, intermediate beaches commonly with a single or multiple bar features are linked with the generation of beach rip currents. When the beach is fully dissipative or reflective, the normal beach rip (central rip) currents are likely to disappear. On the other hand, when natural or man-made structures like headlands, reefs or groynes are present, the surf zone currents are deflected seaward along the sides of the obstacles as topographically-controlled seaward flows of water known as topographic rips or headland rips (Short, 1985). These headland rips are the product of both embayment characteristics and wave height. An increase in wave height re-adjusts the morphological pattern of rip channels and limits the formation of rip currents.

The behaviour of current circulation in embayed beaches has been described by the embayment scaling parameter, δ. However, both approaches still have limitations to predict the headland impact on the surf zone current circulation. Firstly, wave angle is one of the important factors that may contribute to the initiation of rip currents. Although rip currents are normally related to the shore normal waves, oblique waves may change the shape of the rip channel patterns thus limiting the generation of a rip current system. This study for instance, has shown the generation of rip channels both for shore normal ($\theta = 0^0$) and oblique wave ($\theta = 5^0$) under low wave energy conditions (see **Figures 4-4 c and d**, respectively). In both cases, the number of rips present in longer embayments (L=1500 m) for shore normal waves is greater than for oblique waves.

Secondly, initial beach curvature may influence the dynamic circulation in embayed beaches. Although there is no cases tested in this study, Castelle and Coco (2012) have shown in their simulations that both beach curvature and directional spreading of waves influence the generation of central rip currents in a shorter embayed beach.

Thirdly, rip current generation in embayed beaches is dynamic and unpredictable due to a variety of processes and a range of complex behaviour. Rip channels can be developed on a time scale of days to weeks (e.g. Gallop et al., 2011: Ojeda et al., 2011; Castelle and Coco, 2012; Ranasinghe et al., 1999; Damgaard et al., 2002) and it is very difficult to reach an equilibrium state. A proper definition of time scale should be taken into account to properly explain the description of the beach circulation types.

4.4 Results and discussions : Effect of non-stationary waves

All model simulation results presented in **Section 4.3** were produced by XBeach models using a stationary wave solver. Stationary wave means wave is uniform and wave energy distribution is constant throughout the time. In XBeach, an innovative instationary wave solver was implemented, taking into account the contribution of regular and irregular wave-group (infragravity) motions. A generation of irregular wave groups in XBeach is constructed from a single summation, random phase model following van Dongeren et. al (2003). A conversion procedure from wave spectra to wave energy and bound long wave was given in the XBeach manual (Roelvink et al., 2009).

A separate set of additional models were developed taking into account the irregular wave-group resolving forcing. Model setups follow similar approaches as outlined in **Section 4.2**, but this time offshore wave boundary condition is applied as time-varying irregular wave-groups. The offshore boundary was set at a wave height (H_{mo}) of 1.4 m. This value corresponds to an offshore wave height when the stationary wave is used, i.e H_{rms}= 1.0 m. The wave directional spreading of 14^0 was used in all instationary models, which is consistent with the previous model setup that is applied for stationary wave. This directional spreading value was used in several hypothetical embayed beach models (e.g. Castelle and Coco, 2012).

Since a non-stationary wave computation is applied, the breaking wave formulation of Roelvink (1993) is used with the breaking coefficient (γ) of 0.55, a common value for irregular waves. This breaking formula accounts for the slowly-varying dissipation as a function of the local, slowly varying wave-group energy. Short wave friction (f_w) with a value of 0.03 was used. A sediment transport model of Van thiel-Van Rijn was applied. Bed level is then updated after 12 hours. Morphological acceleration factor of 5 was used to speed-up computational time, similar to the case when the stationary wave is applied.

4.4.1 General findings

The morphodynamic of embayed beaches under the influence of wave-group forcing was investigated. Shoal bars and rip channel patterns were produced by the XBeach model for the case when an instationary wave is applied. There are some differences between model's results simulated by instationary and stationary waves models. An obvious difference is that the time scale of the nearshore morphological bed evolution of the instationary case is significantly faster than the evolution time scale of a stationary case.

For the instationary case, rip currents and associated rip channels that are developed in the middle embayment disappear quicker than for the stationary case. As a consequence, smooth bathymetry is achieved at the end of model simulation. Reniers et al. (2004) found a rapid change in morphological evolution in their models when the wave-group forcing is taken into account. The time scale of the morphological evolution was significantly faster in their models compared to the morphological evolution time scale obtained by Damgaard et al., (2002). The reason was due to the presence of additional cross-shore sediment transport contributions associated with both wave asymmetry and wave-induced mass flux.

In our model cases, we believe that the reason of the morphological evolution time scale difference between the stationary and instationary cases is due to the fact that XBeach uses different breaking wave formulations for different wave solvers. For the stationary cases, a parametric model of Baldock (1998) is used to compute the wave energy dissipation due to breaking. Baldock's model resolves the macro scale effects of the wave breaking processes in terms of averaged loss of energy of the ordered wave motion (Battjes and Janssen, 2008), thereby preferable for the stationary wave computation. Indeed, it requires less computational time than for the non-stationary wave computation.

On the other hand, wave energy dissipation due to breaking for the instationary cases is computed by using the probabilistic model of Roelvink (1993). Non-stationary waves mean random waves are imposed over the time and are realistically described by the probabilistic model as such introduced by Roelvink (1993). This method is particularly useful if a detailed wave height distribution is required in the inner surfzone (Baldock, 1998).However, this probabilistic model is limited by the fact that wave-wave interactions in the nearshore and slowly varying water level fluctuations must be minimal (Hamm et al., 1993).

Another possible explanation of the rapid changes in morphological evolution in the model results which are modelled by an instationary wave is due to the presence of energetic surf-beats (infragravity waves) in the nearshore zone. The presence of energetic surf-beats and non-linear effects in the inner surf zone greatly increase the surf zone longshore velocity variance (Guza and Thornton, 1985), thereby can speed-up the morphological bed evolution. Both laboratory study (van Thiel de Vries et al., 2006) and field measurements (Wright et al., 1982, amongst others; Aucan et al., 2008) support that infragravity energy dominates the wave energy spectrum in very shallow water and the swash zone.

Furthermore, each of these instationary cases by essence necessitates an instationary model which allows water waves to occasionally reach places which are inaccessible in the stationary model, due to variations in water level induced by bound long waves. In the instationary case, the point of maximum average water level set-down is closer to shore than in the stationary case. This is directly related to the fact that the position of the breakpoint of the short-waves is closer to shore in the instationary case. The variations of water level near the shoreline between the stationary and instationary models are significant. Inconsistent water levels in the inner surf zone generated by a model when an instationary wave is used enhance surf zone current velocities. This may contribute to the rapid changes of the nearshore seabed evolution.

Discussions on embayed beach morphodynamics under wave-group forcing are presented in **Section 4.4.2. Section 4.4.3** is included to discuss the model results which are related to the effect of various wave directional spreading coefficients on rip current circulation under the wave-group forcing.

4.4.2 Evolution of sand bars and rip channels under wave-group forcing

A set of model simulations were carried out, considering both a shore normal wave ($\theta=0^0$) and an oblique wave ($\theta=5^0$) that to be applied on a medium alongshore scale embayment (L=1500 m).

Figure 4-10 (upper panels) shows the model results of evolution of mean bed level changes produced by XBeach when an instationary wave is applied for a shore normal wave incident. Rip channels develop in the middle of embayment, despite the bed evolution speed is significantly faster than for the model when a stationary wave is applied. An alongshore sand bar starts to evolve at the beginning of day three creating non-uniform shoal bars and rip channels in the middle of the embayment. However, the presence of shoal bars does not remain long as the shoal bars quickly weld to the shore due to the energetic surf-beat. As a result, the shoal bars are hardly visible after the fourth day. The rip current circulations appear to remain in the middle of the embayment until the fifth day. This phenomenon seems to contradict the results obtained from the model with a stationary wave. When a stationary wave is applied, the shoal bars and rip channel patterns in the middle of embayment are clearly visible. The rip circulations remain visible until the end of simulation day (see **Figure 4-5** for comparison).

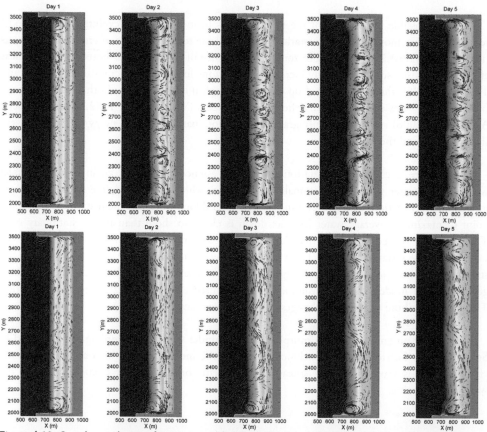

Figure 4-10. Snapshot evolution of time-averaged bed level changes of $H_{rms}=1.0$ m under a shore normal wave-group (upper panels) and an oblique wave-group (lower panels) for an alongshore embayment length of 1500 m.

In the case when the wave are obliquely approached the shore, the model results when the instationary wave is applied show different behaviour from the model results when a stationary wave is used. **Figure 4-10 (lower panels)** shows the snapshot results of the bed level changes with the application of instationary waves when the wave approaches the shore obliquely. The alongshore sand bar merges to the shore quicker than for the case when the stationary wave is applied. The generation of middle rip channel is hardly seen and we suspect this might be due to the wave directional spreading effect. For this simulation, the wave directional spreading of 14^0 was used. This value indicates a quite narrow distribution of wave energy spectrum. A narrow banded spectrum means less alongshore currents can be generated.

4.4.3 Effect of wave directional spreading

Knowledge of directional spectra regarding their evolution in shallow water is important to coastal engineers. Neglecting the wave directional spreading results in over-prediction of significant wave heights and flow velocities. In embayed beaches, directional spreading may cause shadowing effect thus influence current velocities near the headland structure. Castelle and Coco (2012) showed that wave shadowing increases with increasing directional spreading. For a given headland geometry with an alongshore-uniform embayed, the presence of headland rips is favoured by the large directional spreading (Castelle and Coco , 2012).

In this study, we investigated the effect of wave directional spreading on the morphodynamics of embayed beaches under the influence of irregular wave-group forcing. Model simulations incorporate a wide range of directional spreading coefficients ranging from a small directional spreading coefficient (s=1, σ_θ=57.29^0) to a large directional spreading coefficient (s=1000, σ_θ=2.56^0). Small (large) directional spreading coefficient leads to broad (narrow) band wave energy spectrum. In Xbeach, for an irregular instationary wave the frequency distribution of the corresponding energy density is given by a Jonswap spectrum with a cos ^{2s}-directional distribution with 's' coefficient is limited from 1 to 1000. The larger the s, the narrower the energy spectrum.

Figure 4-11 shows model results of various directional spreading coefficients when the incident wave approaches perpendicular to the shore (θ=0^0). The effect of directional spreading can be clearly seen on the morphological shoal bar patterns within the embayment basin. The lesser the directional spreading, the smaller the evolution change in the morphological bed pattern. Large directional spreading induces more alongshore currents which can interact with the nearshore bed. These alongshore currents result from the alongshore variation in wave energy associated with the wave groups. The alongshore currents induce inhomogeneous current circulation patterns in a region between the sand bar and the shoreline, creating rip channels and asymmetrical shoal bar patterns. On the other hand, small directional spreading that creates a narrow-banded spectrum produces less alongshore variations in wave energy. This means the wave energy distribution is mainly produced by the cross-shore standing wave-group only. As a result, the computed bathymetry at the end of day four remains unchanged. (see last panel in **Figure 4-11**).

The application of wave directional spreading coefficient on the embayed beach modelling has contributed to the generation of headland rips. The presence of headland rips is favoured by a large directional spreading coefficient. This is consistent with the model results obtained by Castelle and Coco (2012), despite they did not apply wave-group forcing in their modelling study. Large directional spreading causes greater shadowing effects thus leads to the generation of headland rips. The cross-shore currents that flowing along the headland structures for the case of small directional spreading are expected to occur which are caused by the deflection of water waves against the structure, not the shadowing effects (see **Figure 4-11-last panel**).

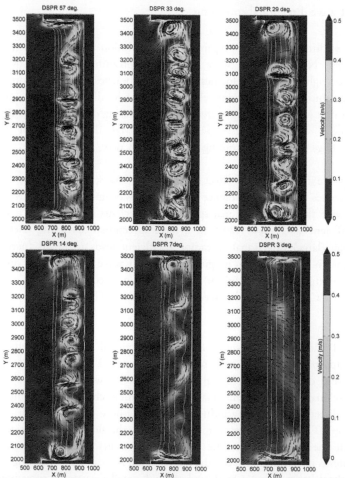

Figure 4-11. Snapshot of time-averaged current magnitudes, superimposed velocity vectors, and bed level contours at t=day 4 for narrow banded directional distribution energy spectrums. The iso-bed contours are plotted at 0.5 m interval from depths 0 to -3 m.

Reneirs et al., (2004) reported that the reason for the observed difference in alongshore separation of the rip channels for the different cases of directional spreading is the underlying quasi-steady circulation pattern on the initial bathymetries (condition before the bed is updated).

The presence of wave-group induces spatially varying mass and momentum fluxes and (alongshore) pressure gradients resulting in an inhomogeneous velocity field. The ensuing circulations depends on the alongshore length scales of the wave-groups and thus the directional spreading of the short waves.

4.5 Mechanism for transporting material out of the surf zone

The accepted view of rip currents was that they are an efficient mechanism for transporting material out of the surf zone. In **Chapter 2**, we did discuss the ability of surf zone currents in transporting sediment away from the beach out-passes the surf zone to the offshore. This traditional paradigm (rip current transports material out of surf zone) has been challenged by recent field (MacMahan et al., 2010), numerical (Reniers et al., 2009) and laboratory studies (Castelle et al., 2009). Rip current circulation patterns actually most of the time consist of semi-enclosed vortices that retain suspended materials within the vortex centre and remain within the surf zone. During the numerical and field experiments, about 10 to 20 % of the drifters deployed in the rip currents exit the surf zone per hour (Castelle et al., 2013).

In this study, we employed virtual drifters on two embayed beaches, one with a medium alongshore scale of 1500 m and another one with a short alongshore scale of 500 m. The virtual drifters are placed on an initial fixed bathymetry. The bathymetry was obtained from the previous model runs, i.e. the final bathymetry which was extracted from the model at the last time step. For each simulation, drifters are uniformly seeded at 5 m intervals at 0 m < x < 1500 m for a medium scale embayment, 0 m < x < 500 m for a small scale embayment and 100 m < y < 200 m within the embayment basin. In total, 6363 and 2079 number of drifters were initially placed in the medium and small embayed beach basins, respectively. Drifter trajectories are calculated at every 1 minute using GLM velocities. The drifters movement were observed from t= 0 min to t= 60 min, under two different wave directions i.e $\theta=0^0$ and 5^0 with a constant wave height of 1 m and a peak period of 10 sec. The drifters exit is calculated based on the number of drifter that crosses the headland boundary, i.e at y > 370 m. It is expected that, by this boundary limit drifters may also exceed the surf zone compartment. The outer surf zone boundary is determined based on the cross-shore wave energy that exceeds 10 % of its cross-shore maximum (Reniers et al., 2009).

4.5.1 Medium embayment basin

Figure 4-12 shows the snapshot plots of drifters movement from t= 0 minute to t= 60 minutes for an alongshore scale embayment length of 1500 m when waves arrive the shore obliquely, both for cases when stationary and instationary waves are applied. In general, drifters that exit the embayment basin are significant when the wave approaches the shore obliquely. For a case when a stationary wave is used, drifters start to flow out of the embayment by 3 %, 30 minutes after the initial drifters are seeded (see **Figure 4-12, stationary wave panels**). The number of drifters exit increases by 9 % at t= 40 minutes, followed by 11 % and 13 % at t= 50 and 60 minutes, respectively. The number of drifters that exits the embayment basin are greater for the case when the instationary wave is applied compared to the number of drifters exit for case when the stationary wave is used (see **Figure 4-12, in-stationary wave panels**). About 46 % of drifters exit the embayment by the end of t= 60 min. Most of the drifters seem to be mainly transported out of the basin by a headland rip in the direction of the approaching wave.

Stationary wave

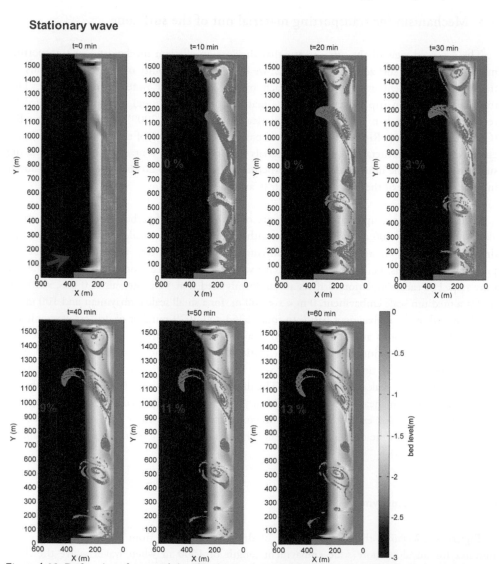

Figure 4-12. Drifter plots of wave with $\theta = 5^0$(inclined red arrow) for a reference case (a) 0 (b) 10, (c) 20, (d) 30, (e) 40, (f) 50 and (g) 60 minutes after virtual drifters seeding in the surf zone. Percentage values in each panel indicate number of drifters that exits the embayment. Alongshore embayment length is 1500 m.

In-stationary wave

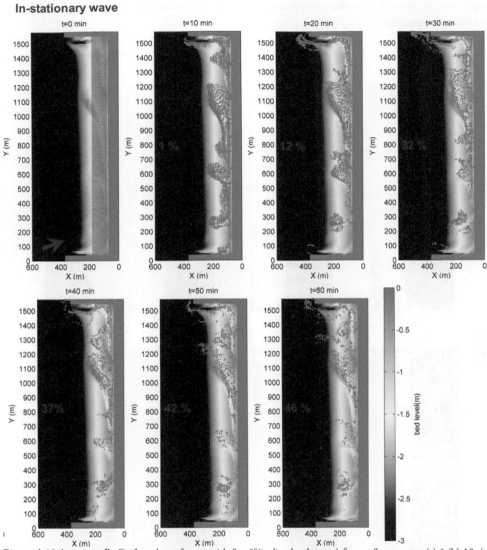

Figure 4-12 (continued). Drifter plots of wave with θ = 5⁰(inclined red arrow) for a reference case (a) 0 (b) 10, (c) 20, (d) 30, (e) 40, (f) 50 and (g) 60 minutes after virtual drifters seeding in the surf zone. Percentage values in each panel indicate number of drifters that exits the embayment. Alongshore embayment length is 1500 m.

Figure 4-13 shows the snapshot plots of drifter's movement from t= 0 minute to t= 60 minutes for an alongshore scale of 1500 m when waves arrive the shore perpendicularly. For this particular case, about 97 % of the drifters remain within the embayment basin. As a result, the computed percentage of drifters exit are zero for all times, regardless of stationary or instationary cases. An obvious difference in both model results is the circulation pattern of the drifters. Model runs with a stationary wave show a semi-enclosed circulation pattern within the rip channels (see **Figure 4-13-stationary wave panels**) while for a model runs with an instationary wave, results show the distribution pattern of drifters are much scattered (see **Figure 4-13 in-stationary wave panels**).

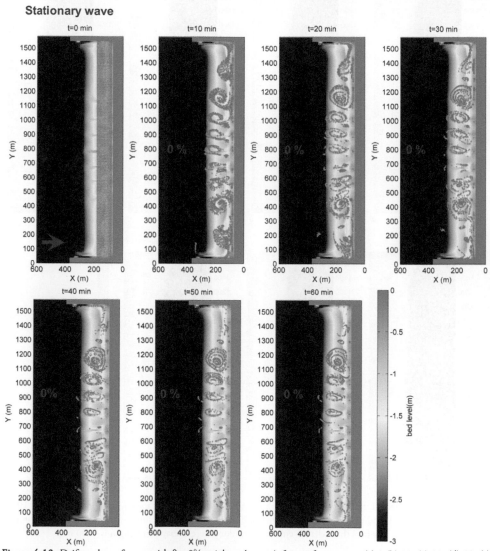

Figure 4-13. Drifter plots of wave with θ = 0⁰(straight red arrow) for a reference case (a) 0 (b) 10, (c) 20, (d) 30, (e) 40, (f) 50 and (g) 60 minutes after virtual drifters seeding in the surf zone. Percentage values in each panel indicate number of drifters that exits the embayment. Alongshore embayment length is 1500 m.

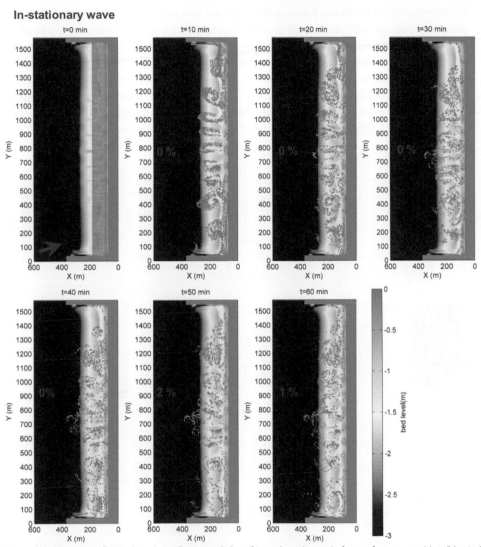

Figure 4-13 (continued). Drifter plots of wave with θ = 0°(straight red arrow) for a reference case (a) 0 (b) 10, (c) 20, (d) 30, (e) 40, (f) 50 and (g) 60 minutes after virtual drifters seeding in the surf zone. Percentage values in each panel indicate number of drifters that exits the embayment. Alongshore embayment length is 1500 m.

4.5.2 Small embayment basin

Figure 4-14 shows the drifter plots in a small embayment with an alongshore length of 500 m for an incoming oblique wave (θ=5°). In both stationary and instatationary cases, drifters are shown to move out of embayment basin and primarily are transported by the headland rips. Results show that model runs with instationary wave produces scattered drifter positions. This is similar to the model results for the medium scale embayment i.e L=1500 m.

The movement of drifters exit between the stationary case and instationary case is different. In the stationary case, the drifters are most likely transported farther away to offshore by a headland rip. The percentages of drifters exit increase over the course of simulation. By the end of t=60 min, about 45 % of drifters exit the embayment basin. Some of the drifters are remain to circulate in the middle of embayment due to the presence of central rip channel. Although in the in-stationary cases the drifters seem to move out of the embayment basin by the headland rip, yet the drifters tend to move just around the headland tip and most unlikely travel farther away to offshore. Some of the drifters are largely scattered in the middle of the embayment basin.

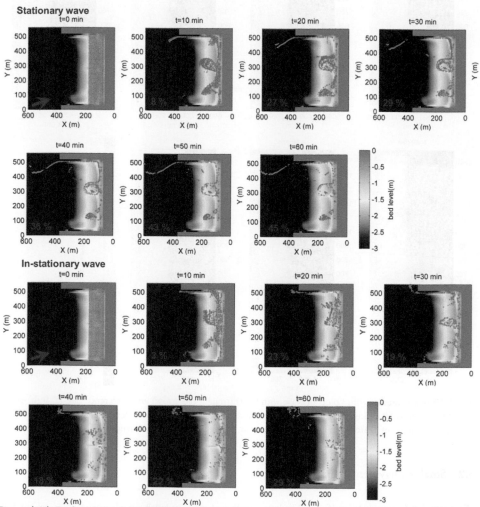

Figure 4-14. Drifter plots of waves with $\theta = 5^0$ for a reference case (a) 0 (b) 10, (c) 20, (d) 30, (e) 40, (f) 50 and (g) 60 minutes after virtual drifters seeding in the surf zone. Alongshore embayment length is 500 m. Percentage values under each panel shows exited drifter.

The computed drifters that exit the small embayment basin for a case when the incident wave approaches the shore perpendicularly are less than for the case case when the incident wave approaches the shore obliquely (see **Figure 4-15**). The drifters exit is hardly seen in the model for the case when stationary wave is applied. In the case when instationary wave is used, the drifters although transport beyond the embayment basin, yet remain to circulate near the headland tip just outside the basin. Headland rips responsible to carry the drifters away from the basin. Most likely, the drifters do not bypass the headland structure.

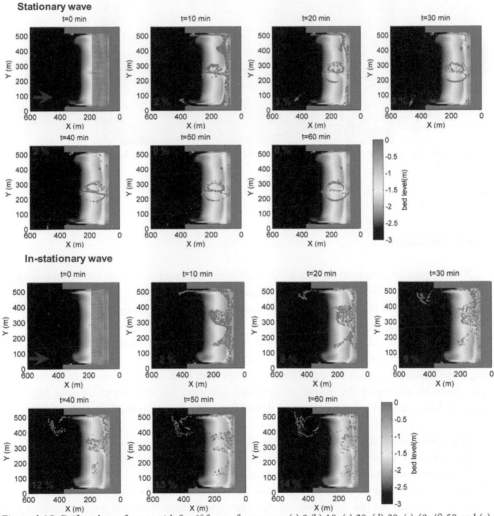

Figure 4-15. Drifter plots of waves with θ = 0⁰ for a reference case (a) 0 (b) 10, (c) 20, (d) 30, (e) 40, (f) 50 and (g) 60 minutes after virtual drifters seeding in the surf zone. Alongshore embayment length is 500 m. Percentage values under each panel shows exited drifter.

Figure 4-16 shows the snapshot of the headland rips that exist on a real embayed beach. The present of this headland rip proves that it can be a medium of the transport mechanism in exchanging sediment between the surf zone and the shelf.

Figure 4-16. Photo of a headland rip extending beyond the surf zone (photo by Andrew Short). Adapted from Castelle and Coco (2013).

4.6 Conclusions

The impact of structural headlands on surf zone current circulation and morphology of several embayed beaches was investigated. Empirical models were used to characterize embayed beach circulation under different wave conditions and different alongshore scale embayment lengths. The wave-group forcing was included in the model simulations and model's results were compared to the model cases without the inclusion of wave-group forcing. Virtual drifters were employed to prove the hypothesis that the surf zone currents are responsible for transporting suspended materials outside the embayment. Models with medium and small embayment basins were build to investigate the impact of structural headland on the cross-shore sand bypassing process. Some detailed conclusions can be drawn from this chapter as the followings:

1. The embayment scaling parameter (δ) that was used as a baseline to describe the characteristics of morphodynamics of embayed beaches proved to synchronize with the development of predicted current circulation patterns.
2. In this modelling study, surf zone currents such as headland rips and central beach rips are proven existed on an embayed beach. These surf zone currents are consistent with observations in several real embayed beaches.
3. Additionally, the transition between the transitional beach state to a cellular beach state has been proven under increasing wave energy for a longer scale embayment.
4. The inclusion of extreme wave events in this modelling study showed a different behaviour of surf zone current circulation pattern which has not been seen in the field as reported in

literature due to several factors i.e nearshore and adjacent embayment topography and breaking wave conditions. Although a significant rip channel in the middle of embayment exists, this is only due to the increasing strength of the existing current.

5. Wave breaking condition plays an important role in the generation of large scale rip current.

6. Further numerical investigation should be carried out to determine the related processes that drive the formation of megarip current.

7. The inclusion of irregular wave-groups (instationary waves) in this modelling study showed their presence cause rapid changes in bed level. This might be due to different energy dissipation in a wave breaking formulation and the presence of energetic surf-beats in the inner nearshore zone.

8. Irregular wave-groups tend to smoothen the final nearshore bathymetry. The alongshore sand bar is pushed towards the shore by the standing cross-shore waves diminishing the existence of central rip channels.

9. Wave directional spreading contributes to the morphodynamic of embayed beaches. The larger the directional spreading, the greater the change in morphological bed pattern.

10. Wave directional spreading causes wave shadowing effect. The presence of   headland rips are favoured by a large directional spreading.

11. Floating materials appear to exit an embayment basin mainly by the   headland rips, particularly when an incident wave approaches the shore   obliquely.

12. For a small scale embayment, the drifters exit the embayment basin through the headland rips while for a medium scale embayment, the surf zone rip currents have considerable space to circulate within the embayment and deflected by the headland rips.

13. When modelled with an irregular wave-group forcing, drifters are dispersedly   distributed in the nearshore zone.

Although results presented in this study are purely hypothetical, it replicates the real phenomena observed in nature (e.g Loureiro et al., 2012, Gallop et al., 2011, and Short and Masselink, 1999). Future study is recommended to include the storm wave-grouping to observe the effects of real storm waves to the development of surf zone current circulation in small scale embayed beaches.

References

Ab Razak, M.S., Dastgheib, A., and Roelvink, D. (2013). Sand bypassing and shoreline evolution near coastal structure comparing analytical solution and XBeach numerical modelling, *Journal of Coastal Research*, SI No. 65: 2038-2044.

Aucan, J., Pequignet, A.C., Vetter, O.J., Becker, J.M. and Merrifield, M.A. (2008). Wave transformation and setup across Ipan Reef, Guam during tropical storm Man-Yi, *Ocean Sciences Meeting*, Orlando, Florida.

Battjes, J. A. and Janssen T. T. (2008). Random wave breaking models - History and discussion, *Coastal Engineering 2008*, 25:37.

Baldock, T.E., Holmes, P., Bunker, S. and Van Weert, P. (1998). Cross-shore hydrodynamics within an unsaturated surf zone. *Coastal Engineering* 34: 173- 196.

Bowen, A.J. (1969). Rip currents. 1. Theoretical investigation. *Journal of Geophysical Research*, 74( ): 5467-5478.

Calvette, D., Coco, G., Falques, A., and Dodd, N. (2007). (Un) predictability in rip channel systems, *Geophysical Research Letter*, 34(): L05605.

Coutts-Smith (2004) The significant of megarips along an embayed coastline. PhD thesis University of Sydney, Australia. 221p

Castelle, B. Reneirs, A. and MacMahan, J. (2013). Numerical modelling of surfzone retention in rip current system: On the impact of the surfzone sandbar morphology, *Coastal Dynamics 2013*: 295-304.

Castelle, B. and Coco, G. (2013). Surf zone flushing on embayed beaches, *Geophysical Research Letters*, 40: 1–5. doi:10.1002/grl.50485.

Castelle, B. and Coco, G. (2012). The morphodynamics of rip channels on embayed beaches, *Continental Shelf Research*, 43( ): 10-23.

Callaghan, D.P, Ranasinghe, R., and Roelvink, D. (2013). Probabilistic estimation of storm erosion using analytical, semi-empirical, and process based storm erosion models. *Coastal Engineering*, 82( ): 64-75.

Chappell, J. and Wright, L.D., (1978). Surf zone resonance and coupled morphology. *Proceedings of the 16th International Conference on Coastal Engineering*, Hamburg, Germany. ASCE,pp. 1359-1377.

Cowell, P.J. (1975). Morphodynamic aspects of the interactions between incoming waves and bed topography at Palm Beach, N.S.W. Unpublished Thesis, Department of Geography, University of Sydney, 179p.

Dalrymple, R.A., Mac Mahan, J.H., Reniers, A.J.H.M., and Nelko, V. (2011). Rip currents, *Annual Review of Fluid Mechanics*, 43(): 551-581.

Damgaard, J., Dodd, N., Hall, L., and Chesher, T. (2002). Morphodynamic modelling of rip channel growth. *Coastal Engineering*, 45(): 199-221

Guza, R.T. and Thornton, E.B. (1985). Observations of surf beat. *Journal of Geophysical Research*, 90(C4): 3161-3172.

Gallop, S.L., Bryan, K.R, Coco. G., and Stephens, S.A. (2011). Storm-driven changes in rip channel patterns on an embayed beach, *Geomorphology*, 127(): 179-188.

Gourlay, M.R., (1968). Beach and Dune Erosion Tests. Delft Hydraulics Laboratory, Report M935/M936.

Hamm, L., Madsen, P.A., and Peregrine, D.H. (1993). Wave transformation in the nearshore: a review, *Coastal Engineering*, 21 (1993): 5–39.

Holman, R., Symonds, G., Thornton, E.B., and Ranasinghe, R. (2006). Rip spacing and persistence on an embayed beach. *Journal of Geophysical Research*, 111(): C01006.

Huntley, D.A., Hendry M.D., Haines, J., and Greenidge, B. (1988). Waves and rip currents on a Caribbean Pocket Beach, Jamaica. *Journal of Coastal Research*, 4(1): 69-79.

Komar, P.D. (1998). Beach Processes and Sedimentation, 2nd ed. Upper Saddle River, New Jersey: Prentice Hall, 544p.

Lees, B.G. (1977). The effects of compartmentalization on beach processes and forms: Cronulia, N.S.W. Unpublished Thesis, Department of Geography, University of Sydney, 179p.

Loureiro, C. Fereirra, O., and Copper, J.G. (2012). Extreme erosion on high energy embayed beaches; Influence of megarips and storm grouping, *Geomorphology*, 139-140(): 155-171.

MacMahan, J.H., Brown, J.W., Brown, J.A., Thronton, E.B., Reniers, A.J.H.M., Stanton, T.P., Henriquez, M., Gallagher, E.L., Morrison, J., Austin, M.J., Scott, T.M. and Senechal, N., (2010). Mean Lagrangian flow behavior on an open coast rip-channeled beach. A new perspective, *Marine Geology*, 1-4(): 1-15.

MacMahan, J.H., Thornton, E.B., Reniers, A.J.H.M, Stanton, T.P., and Symonds, G. (2008). Low energy rip currents associated with small bathymetric variations, *Marine Geology*, 255(): 156-164.

McKenzie, P. (1958). Rip current systems. *Journal of Geology*, 66( ): 103-113.

Munk, W.H. (1949) Surf beats. *Eos Trans. Amer. Geophys. Union*, 30: 849-854.

Ojeda, E., Guillén, J., and Ribas, F. (2011). Dynamics of single-barred embayed beaches, *Marine Geology*, 280(): 76-90.

Ranasinghe, R., Symonds, G., and Holman, R. (1999). Quantitative characterization of rip currents via video imaging, *Proceedings of Coastal Sediments '99*, New York, USA, pp. 987 - 1002.

Reniers, A.J.H.M., Roelvink, J.A., and Thornton, E.B. (2004), Morphodynamic modeling of an embayed beach under wave group forcing, *J. Geophys. Res.*, 109, doi:10.1029/2002JC001586.

Reniers, A.J.H.M., Mac Mahan, J. H., Thornton, E.B., Stanton, T.P. Henriquez, M., Brown, J. W., Brown, J. A. and Gallagher, E. (2009). Surf zone retention on a rip-channeled beach, *J. Geophys. Res.*, 114, doi:10.1029/2008JC005153.

Reniers, A. J. H. M., MacMahan, J.H., Beron-Vera, F.J., and Olascoaga, M.J. (2010), Rip-current pulses tied to Lagrangian coherent structures, *Geophys. Res. Lett.,* 37, DOI:10.1029/2009GL041443.

Roelvink, D., Reneir, A.J.H.M., van Dongeren, A., van Thiel de Vries, J., McCall., R., and Lescinski, J. (2009). Modelling storm impacts on beaches, dunes and barrier islands. *Coastal Engineering,* 56,11-12():1133-52.

Roelvink, J.A., Reniers, A.J.H.M., van Dongeren, A., Vries, J.T., Lescinski, J., and McCall, R. (2010). XBeach model description and manual. (UNESCO-IHE Institute for Water Education, Deltares, Delft University Technology).

Roelvink, J.A. (1993). Surf beat and its effect on cross-shore profiles. PhD Thesis, Technical University of Delft, the Netherlands, 150 p.

Shepard, F.P., and Inman, D.L. (1950). Nearshore water circulation related to bottom topography and wave refraction. *Transaction of the American Geophysical Union,* 31(): 196-212.

Short, A.D. (1985). Rip current type, spacing and persistence, Narrabeen Beach, Australia, *Marine geology,* 65(): 47-71.

Short, A. D. and Masselink, G. (1999). Embayed and structurally controlled beaches, In: Short, A.D.(ed)., *Handbooks of Beach and Shoreface Hydrodynamics.* Chicester: John Wiley & Sons. pp. 230-249

Short, A.D. (2007). Australian rip systems: friend or foe? *Journal of Coastal Research,* SI 50: 7-11.

Short, A.D. (2010). Role of geological inheritance in Australian beach morphodynamics. *Coastal Engineering,* 57(): 92-97.

Silva, R., Baquerizo, A., Loasada, M.A., and Mendoza, E. (2010). Hydrodynamic of a headland-bay beach-Nearshore current circulation, *Coastal Engineering,* 57(): 160-175.

Splinter, K.D., Carley, J.T., Golshani, A., and Tomlinson, R. (2013). A relationship to describe the cumulative impact of storm clusters on beach erosion. *Coastal Engineering,*83(): 49-55.

Spydell, M., Feddersen, F., and Guza, R.T. (2007). Observing surf-zone dispersion with drifters, *J. Phys. Oceanogr.,* 37(): 2920–2939.

Symonds, G., Holman, R.A., and Bruno, B. (1997). Rip currents. In: Coastal Dynamics '97, Thorton, E.B. (ed.), *American Society of Civil Engineering,* Reston, VA, pp:584-593.

Tucker, M.J. (1950) Surf beats: Sea waves of 1 to 5 minute period. *Proc. Roy. Soc. London, Ser. A,* 202:565-573.

van Dongeren, A., Reniers, A.J.H.M., and Battjes, J.A., (2003). Numerical modeling of infragravity waves response during DELILAH. *Journal of Geophysical Research,* vol. 108(C9): 3288.

van Thiel de Vries, J.S.M., van de Graaff, J., Raubenheimer, B., Reniers, A.J.H.M and Stive, M.J.F. (2006). Modeling inner surf hydrodynamics during storm surges. In: J. Mckee Smith, *Proc. of 30th International Conference on Coastal Engineering.* World Scientific Pub. Co. Inc. , San Diego.

Wright, L.D., Guza, R.T. and Short, A.D. (1982). Dynamics of a high-energy dissipative surf zone. *Marine Geology,* vol. 45(): 41-62.

Wright, L.D. and Short, A.D., (1984). Morphodynamic variability of surf zones and beaches: A synthesis. *Marine Geology,* 56 (1-4): 93-118.

Wright, L., Thom, B., and Chappell, J. (1978). Morphodynamics variability of high-energy beaches. Coastal Engineering Proceedings, 1(16). pp 1180-1194. doi:http://dx.doi.org/10.9753/icce.v16.%p

CHAPTER 5

Embayed beach response due to sand nourishment on the east coast of Malaysia

The Cempedak Bay beach stability assessment was performed by comparing the spatial and temporal pattern of beach variability before and after sand nourishment. The analysis of temporal sand volume patterns shows that the beach has lost about 6 % or 10 000 m³ volume of sand which is equivalent to 4 m³/m per year from the nourishment zone over the 2.5-year monitoring period. The present shoreline recession rate is established to be 1.7 m/year (valid for data set of March 2005 to July 2007). The analysis of seasonal changes is assessed through temporal beach volume patterns, which indicate that shoreline variability can be characterised by an alongshore rhythmic pattern of alternating seasonal behaviour. A simple seasonal transport pattern is proposed to account for alternating erosion and accretion. The temporal distribution pattern of beach level changes reveals the existence of a nodal point around 40 to 50 m offshore, which is influenced by the monsoonal system. The spatial distribution of the beach width indicates that the northern beach area is wider whereas the southern beach area experiences lower beach width, which is coincident with the temporal pattern of sand volume and beach profile changes. A slight beach rotation does exist attributed to a seasonal or periodic shift in wave climate, in particular wave direction. The planform stability of the beach is tricky to determine due to model uncertainties, especially the selection of the diffraction point. Sand can easily bypass the Tanjung Tembeling headland as the headland is reasonably small.

Major parts of this chapter were published in:

(i) Ab Razak, M.S., Roelvink, D. , and Reyns, J. (2013). Beach response due to beach nourishment on the east coast of Malaysia, *Proceeding of the ICE-Maritime Engineering Journal,*66(4):151-174

(ii) Ab Razak, M.S and Roelvink, D. (2011). Determination of equilibrium stages of headland bay beaches: A preliminary case study on the east coast of Malaysia, *Proceeding of the International Conference of River, Coastal and Estuaries Morphodynamics.* Beijing, China

5.1 Introduction

Land based activities and natural physical processes have resulted in significant modifications of the shorelines in many countries, with drastic effects on the coastal geomorphology as well as on the coastal infrastructures. In Malaysia, soft sea defence such as beach nourishment is a common practice to protect the beach from erosion. One of the bay beaches which has been documented to experience erosion is known as the Cempedak bay beach.

Although this beach is classified as stable under the National Coastal Erosion Study (NCES) (Stanley Consultants, 1985), the average retreat rate is estimated to be 0.8 m/year. If protective measures are not taken, the beach will eventually be eroded, and the ocean waves approaching the land will endanger the properties located along the beachfront and in the hinterland. The re-nourishment of sand every five years is a common tradition to provide wider beach area as the Cempedak beach is a source of financial income for the state government of Pahang. It should be noted that tourism had been accorded a high priority by the federal and state government of Pahang. Thus, an assessment is required to inspect the effectiveness of the system.

In 2004, the combined beach nourishment and Pressure Equalization Module (PEM) system have been installed at the Cempedak bay beach. Three years monitoring of beach profile surveys have been conducted between 2005 and 2007. Although the project includes the PEM system, the performance of that vertical beach drainage system is not taken into consideration for this study. Fair assessment is required for instance, comparing beach changes with only sand nourishment and nourishment with the PEM system, which is limited for this scope of project. Therefore, we only consider the response of the beach due to sand nourishment.

The main objective of this chapter is to see how much the profiles vary through erosion and accumulation processes before and after the nourishment period. The investigation focuses on four main categories: distribution pattern of sand volume changes, distribution pattern of beach level changes, and the development of dry beach width and beach rotation. In addition, the empirical parabolic bay model is employed to determine the state of stability of Cempedak bay beach.

5.2 Literature review

5.2.1 Beach nourishment and sediment budget approach

Beach nourishment as defined by National Research Council (1995) is a technique used to restore an eroding or lost beach or to create a new sandy shoreline, which involves the placement of sand fill with or without supporting structures along the shoreline to widen the beach. It is the only management tool, which serves the dual purpose of protecting coastal lands and preserving beach resources. A major question regarding beach nourishment concerns the success of the project.

A sediment budget approach to examine beach behaviour is an accepted practice used to provide a detailed view of changes in landforms [(Komar, 1983) in (Gares, 2006)]. This approach relies on the determining the volumes of sediment added or removed from the beach. When the additions and reductions are known, they can give a picture, whether the system is losing or gaining volume. This approach to monitoring beach nourishment projects would provide information about the redistribution of the fill materials from the nourished beach to adjacent areas. As the beach nourishment is a volume-based technique, the sediment budget approach is reasonable to monitor the project's progress (Gares et al., 2006). Beach profile surveys, satellite images and aerial photographs are such example techniques used to execute sediment budget calculation as well as to monitor the seasonal variation of the shoreline. Cooper et al., (2000) suggested that beach profiles are an important tool for understanding the long term trends of erosion, accretion and predicting the future landforms. Other examples of successful beach nourishment project are given in Saravanan and Chandrasekar (2010) and Kuang et al., (2011).

5.2.2 Equilibrium stages of headland bay beaches

Hsu et al., (2008;2010) classified the stability of bay beaches into three states i.e. static equilibrium state, dynamic equilibrium state, and unstable or natural reshaping state. See example in **Figure 5-1** (Silveira et al., 2010) for better explanation.

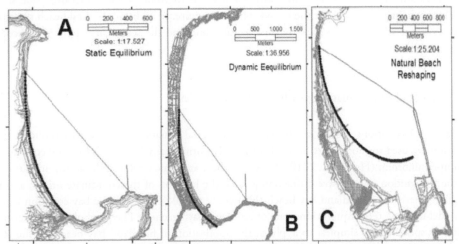

Figure 5-1. Planform stabilities of headland bay beaches. Upper left panel-A (static equilibrium state); Upper right panel-B (dynamic equilibrium state); lower left panel-C (natural beach reshaping). The bold line indicates the static equilibrium shoreline position, described by parabolic bay shape equation. Courtesy of Silveira et al.,(2010).

In a static state, the net longshore sediment transport is approximately zero. It can also be characterized by the presence of storm and /or swell waves from one dominant direction, no further beach changes, and simultaneous wave breaking at every location along the beach. For this case, waves generally diffract around the headland and refract as well as shoal near the beach. The bay shape modelled using the parabolic shape equation (Hsu and Evans 1989) coincides with the shape of beach in static equilibrium Silveira et al., (2010).

On the other hand, beaches in dynamic equilibrium depend on the local sediment budget. It basically differs from the static state equilibrium as follows. Generally, the dynamic beach line is lying seaward of the static beach line whereby the shoreline will migrate landward if sediment supply from up-coast ceases. Littoral transport may be constant, if there is a continuous supply of sand around the up-drift headland. Two factors that influence the curvature shape of this beach are the sand bypassing from the up-drift headland and or sediment input of a river outlet inside the bay (van Rijn, 1998). This process is dictated by the generation of longshore currents thus consequently transporting the sediment when the wave breaks to the shoreline. In contrast to the case of static equilibrium, the shape of the beach does not coincide with the shape modelled by the parabolic equation. The shoreline tends to move toward the equilibrium state if the sediment supply is decreased, (Silveira et al., 2010) causing erosion and damage to infrastructure near the beach area.

The third state called natural beach reshaping or unstable condition normally associated with wave sheltering due to addition or extension of structures on a beach, where a curved planform could result with accretion in the lee accompanying by erosion downdrift. This state can be achievable when there is a coastal engineering structure constructed or placed on the beach such as groynes and breakwaters. For this case, beach tends to erode at the down-drift side where sediment is being transported at the lee side of the structure. In addition, natural beach reshaping also can occur when a beach is in dynamic equilibrium and the sediment supply is reduced or ceases altogether.

5.2.3 Parabolic bay equation of a headland bay model

Empirical equations known as the logarithmic spiral equation and the hyperbolic tangent equation are used to model part or whole of the planform of any bay beach in static equilibrium or non-equilibrium (Hsu et al., 2010). However, there are some obvious limitations, in particular on the recognition of the wave diffraction points, the location of the coordinate origin, and the stability criteria for a headland bay beach (Hsu et al., 2010). The parabolic bay equation is used to model the long term equilibrium state of headland bay beaches. The parabolic bay equation is relatively speaking the simplest to apply in the application of headland bay beaches. The main advantage of this equation is that it takes into account the predominant wave direction and control point at the up-drift headland. Two models which are known as MEPBAY (Klein et al., 2003) and SMC (Gonzales et al., 2007) have been using this equation. However, MEPBAY is the most popular one, easily handled (Klein et al., 2003; Jackson and Cooper 2010; Lausman et al., 2010; Raabe et al., 2010; Silveira et al., 2010) and has been widely applied. Starting from the planform of a headland bay beach on a satellite image, aerial photograph, and map, the model offers an interface that allows users to indicate the relevant control points on the coastline and to trace the complete bay periphery in static equilibrium automatically (Klein et al., 2003). In addition, with little information needed as for input of this model, it is relatively suitable for project evaluation to determine the optimum design options with variable structure configuration. The parabolic bay shape equation for a headland bay beach in static equilibrium reads as follows (**Equation 5.1**)

$$\frac{R}{R_\beta} = C_o + C_1 \left(\frac{\beta}{\theta_n} \right) + C_2 \left(\frac{\beta}{\theta_n} \right)^2$$

(5.1)

where the geometric parameter R, R_β, β, and θ are defines in **Figure 5-2**. The constant value of C_o, C_1, C_2 is generated by regression analysis to fit the peripheries of the 27 prototypes and model bay, differ with reference angle β and can be expressed by fourth-order polynomials as written in **Equation 5.2** (Hsu and Evans, 1989).

$C_o = 0.0707 - 0.0047\beta + 0.000349\beta^2 - 0.00000875\beta^3 + 0.00000004765\beta^4$
$C_1 = 0.9536 - 0.0078\beta + 0.00004879\beta^2 - 0.0000182\beta^3 + 0.000001281\beta^4$
$C_2 = 0.0214 - 0.0078\beta + 0.0003004\beta^2 - 0.00001183\beta^3 + 0.00000009343\beta^4$

(5-2)

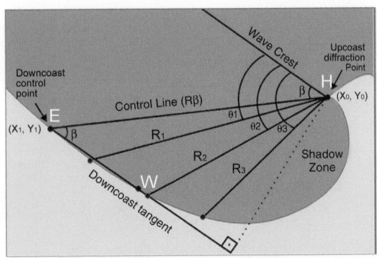

Figure 5-2. Definition sketch for the parabolic bay shape model showing major physical parameters. Point H is the updrift control point, W is the downdrift tangent control point, and E is the downcoast control point. Modified from Hsu and Evans (1989).

5.3 Material and methods

5.3.1 Description of study area

The coasts of Malaysia have a combined length of about 4809 km. About 1415 km (29%) of this 4809 km coastline is subjected to erosion of various degrees of severity (Tan et al., 2005). Erosion may be amplified during monsoon periods when high water levels associated with the seasonal storms result in waves breaking directly against the scarp, causing loss of beach material. Although some of this material might be returned to the shore by swells after the monsoon, the quantity returned is normally less than the amount moved offshore and hence the net result is erosion.

Control of erosion has become an important economic and social need in Malaysia. Specifically, about 73.4% or 52.1 km of the total length of coastline in Pahang (the biggest state in peninsular Malaysia) is in the long-term erosional state (Ghazali, 2006). The study area named Cempedak Bay beach, as shown in **Figure 5-3**, is located on the east coast of Peninsular Malaysia. Orientated at 355°, it is one of the headland bay beaches which have experienced steady erosion problems over the years. It is a pocket beach between the granite headlands of Tanjung Pelindung Tengah and Tanjung Tembeling, with the Cempedak river draining into the northern end of the bay (**Figure 5-4**). This river drains Bukit Pelindung and discharges some sediment and moderately polluted water from developed areas within its catchment.

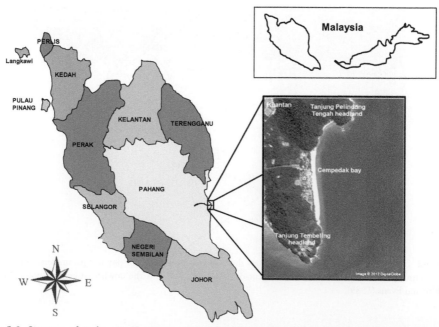

Figure 5-3. Site map of study area. (Satellite image on the right is adapted from Google Earth version 6; image at 2012© DigitalGlobe)

The bay has been fully developed for public recreation as well as for local and international tourism. The public recreational areas occupy the northern part of the beach and the Sheraton and Hyatt Hotels are on the southern part. The beach is about 1.1 km long, swash aligned and reflective. The beach morphology in the undisturbed areas shows two levels of beach cusp due to the differences in the sea level and wave energy of the two monsoons. The beach sands are yellowish and coarse-grained, reflecting the nearby source of alongshore sands from the eroding headlands.

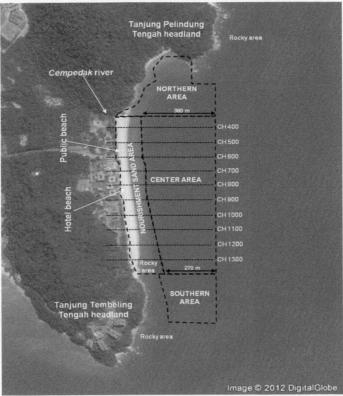

Figure 5-4. Location of sand nourishment area and expected sand distribution areas and profile transects at the Cempedak Bay beach. Background image represents Cempedak Bay beach of 2006. (Satellite image on the right is adapted from Google Earth version 6; image at 2012© DigitalGlobe)

Cempedak Bay beach has a history of erosion. The beach area has undergone slow and steady erosion that has resulted in the narrowing of the beach area, which has had adverse effects on the recreational and tourist activity in this area. Cempedak Bay is a sandy beach with an original slope of 1:60. This stretch of beach has a narrow and rather steep near-shore and has suffered steady erosion over the years. It has been documented that the shoreline retreat is approximately 0.8 m/year based on the National Coastal Erosion Study of 1985 (Stanley Consultants, 1985). The steep near-shore area is also prone to high wave activity that results in strong nearshore currents which may endanger swimmers. From the previous report produced by the Department of Drainage and Irrigation Malaysia, this scenario may happen as a result of several factors: (a) losses of the sand due to interruption of longshore transport on the updrift side; (b) reduction of sediment source; and (c) storm surges. Meanwhile, the hookshape bay with its coarse sand and relatively steep slope are caused by the turbulence resulting from its swash-alignment and combined wave diffraction–refraction that washes down the rocky headlands of Tanjung Pelindung and Tanjung Tembeling. On the other hand, the beach also consists of two separated sand fractions: one at the berm or upper part (coarse sand) of the foreshore, and one below (fine sand); this may characterise the beach where the higher zone is dominated by breaking waves and the lower zone by shoaling waves. The beach was nourished in 2004 when 177, 000 m³ of sand was placed on the beach. The nourishment scheme started in May 2004 and ended in July 2004.

Based on the historical record (DID, 2007), the normal lifetime of the traditional beach nourishment at Cempedak Bay is approximately 5 years, and might be shorter than the normal period due to the impact of high-energy events and strong hydrodynamics, especially during the north-east monsoon (November to March). For this reason, the Malaysian government has implemented a new strategy to slow down the erosion in which RM15 400 000.00 (approximately £ 3.2 million) has been spent on a new system for protection of coastal erosion, which is called the PEM, in combination with sand nourishment (**Figure 5-4**). The project was started in 2003 and lasted until the end of 2007. For a clear view, satellite images of the years 2003, 2006 and 2007 are presented in **Figure 5-5** to show the development of the beach at Cempedak Bay. However, the present study mainly focused on the response of the beach to sand nourishment. The role and efficiency of the PEM system were neglected and are not included in the discussion herein.

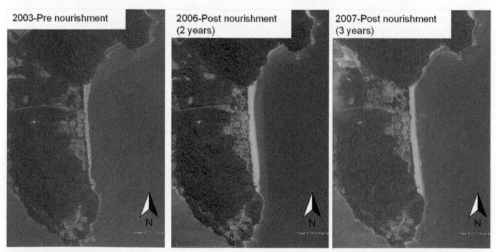

Figure 5-5. Historical development of the Cempedak Bay beach (Image courtesy of Google Earth version 6; image © 2012 DigitalGlobe)

5.3.2 Hydrodynamic and sediment transport condition

For the Pahang coast, the available measured wave data were taken from the wave information of Synoptic Shipboard Meteorological Observation (SSMO) data of waves off the east coast. The SSMO wave data were used to generate the offshore annual and seasonal wave roses as presented in **Figure 5-6**. Wave and wind data were obtained from the Coastal Engineering Division of the Department of Drainage and Irrigation, Malaysia. During the north-east monsoon (November to March), the wind direction is from the north/north-east, whereas, in the south-west monsoon (April to October), the wind blows mostly from the south-west/south/south-east direction. The highest mean wind speed is recorded in December with the average wind speed higher than 8 m/s. The tabulated offshore wind is given in **Table 5-1**. Seasonal variation influences the morphological changes of nourished beaches (e.g. Norcross et al., (2002) and Yates et al., (2009)). This was a result of higher wave crests during the northeast monsoon time and smaller wave crest during the southwest monsoon time. Likewise, Cempedak Bay experiences seasonal beach crest variations due to variation in wave heights for each season.

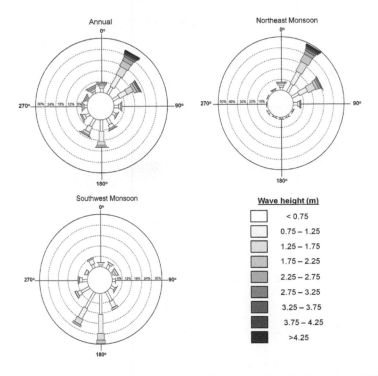

Figure 5-6. Offshore wave roses for the east coast of Malaysia based on the British Meteorological Office (BMO) UK from 1949-1983. The northeast monsoon is from November to March and the southwest monsoon is from April to October.

Table 5-1. Offshore wind for the east coast of Malaysia covering period 1.8.1991 till 31.7.1992, 1998 and 1999.Mean: average wind speed (m/s);Max: maximum wind speed (m/s), Dir (%): percentage of time.

Direction	North			North-East			East			South-East			South			South-West			West			North-West		
Months	Mean	Max	Dir (%)	Mean	Max	Dir (%)	Mean	Max	Dir (%)	Mean	Max	Dir (%)	Mean	Max	Dir (%)	Mean	Max	Dir (%)	Mean	Max	Dir (%)	Mean	Max	Dir (%)
Jan	5	8	1.6	6	11	6.5	3	7	0.2	1	2	0.1	0	0	0.0	0	0	0.0	1	1	0.0	3	5	0.0
Feb	6	10	1.6	5	11	5.3	3	9	0.6	2	4	0.2	0	0	0.0	0	0	0.0	0	0	0.0	0	0	0.0
Mar	2	5	0.5	4	9	4.9	3	7	2.2	2	4	0.5	2	4	0.2	1	3	0.1	1	1	0.0	1	1	0.0
Apr	2	6	0.3	4	10	2.1	3	5	1.9	2	6	1.9	3	6	1.4	3	8	0.3	2	3	0.1	2	4	0.1
May	1	1	0.1	3	6	0.8	3	5	1.6	3	6	1.7	3	7	2.8	3	6	1.4	1	2	0.1	2	2	0.1
Jun	2	3	0.1	3	4	0.1	2	4	0.3	3	6	1.3	4	8	4.5	4	7	1.9	1	1	0.0	0	0	0.0
Jul	1	2	0.1	1	1	0.0	3	5	0.4	3	7	0.7	5	10	4.9	4	10	2.1	2	3	0.2	2	2	0.1
Aug	1	1	0.0	1	2	0.0	1	2	0.1	3	5	0.4	5	10	5.3	5	9	2.4	1	2	0.1	1	2	0.1
Sep	2	2	0.1	2	3	0.1	1	3	0.2	3	7	0.8	4	10	4.6	4	8	2.3	2	4	0.1	1	2	0.1
Oct	2	6	0.8	2	7	0.9	2	5	0.9	2	4	0.9	3	6	1.8	3	6	1.4	3	6	1.0	3	6	0.8
Nov	4	9	2.5	5	10	3.1	2	4	0.5	2	3	0.3	2	3	0.1	1	4	0.3	2	4	0.7	2	5	0.8
Dec	7	14	3.1	7	13	3.9	4	9	0.5	3	4	0.2	6	7	0.1	5	5	0.0	4	6	0.1	5	8	0.5

For example, during the northeast monsoon the waves reach up to 4 m, whereas in the southwest monsoon the maximum waves appear to be around 2 m (**Figure 5-6**). This difference in wave heights could be the process responsible for the different seasonal beach crest heights seen in the profiles as later discussed in Section 4. The mean high water is 2.8 m chart datum (CD) and the type of tide is diurnal. No sediment input was transported from the sea into the pocket bay (DID, 2007). The native sand at Cempedak Bay consists of two separated sand fractions, one at the upper part of the foreshore (swash zone), which is above mean sea level (0.05m land survey datum (LSD)), and the other one at the lower part which is below mean sea level. The sand at the upper part consists of coarse sand whereas the lower part is fine sand. For the upper part, the neap and spring tide samples of coarse sand are within the range 0.33 to 0.85 mm and 0.60 to 1.00 mm, respectively. For the lower part, the neap and spring tide samples of fine sand are within the ranges 0.15 to 0.20 mm and 0.14 to 0.18 mm, respectively. The sediment samples were collected in the areas of public beach and hotel beach as identified in **Figure 5-4**. The transport of sediment is mainly from the north to the south during the monsoon time and vice versa the next season.

5.3.3 Beach nourishment

The project area was evaluated for the first time in March 2003; this was pre-nourishment. The nourishment was started in May 2004 and ended in July 2004 where Cempedak Bay beach was nourished with 177 000 m³ of sand. The monitoring surveys were further conducted three times a year after the nourishment period, namely in October, March and July, which covered the period of the north-east monsoon and south-west monsoon. The monitoring period ended in July 2007. The nourished beach is 1100 m long, and the sand was placed from the concrete wall and 100 m out to the sea as depicted in Figure 4. **Table 5-2** shows the summary of design size ranges for borrowed sand which was dredged from the seabed offshore with the sand source located about 5 km from the shoreline and at a water depth of approximately 14 m.

Table 5-2: Plan view of location for design size sand. Mean sea level (MSL) is 0.05 m LSD.

Beach area	Profile	Upper beach face (above MSL)	Lower beach face (below MSL)
Public beach	Profile 400	D_{50} = 0.820 mm	D_{50} = 0.280 mm
	Profile 500	D_{50} = 0.820 mm	D_{50} = 0.280 mm
	Profile 600	D_{50} = 0.820 mm	D_{50} = 0.280 mm
Hyaat Hotel beach area	Profile 700	D_{50} = 1.000 mm	D_{50} = 0.330 mm
	Profile 800	D_{50} = 1.000 mm	D_{50} = 0.330 mm
	Profile 900	D_{50} = 1.000 mm	D_{50} = 0.330 mm
Faber/Sheraton Hotel beach area	Profile 1000	D_{50} = 1.000 mm	D_{50} = 0.330 mm
	Profile 1100	D_{50} = 1.000 mm	D_{50} = 0.330 mm
	Profile 1200	D_{50} = 0.820 mm	D_{50} = 0.280 mm
	Profile 1300	D_{50} = 0.820 mm	D_{50} = 0.280 mm
	Profile 1400	D_{50} = 0.820 mm	D_{50} = 0.280 mm

* LSD refers to land survey datum

5.3.4 Beach stability assessment

A series of monitoring profile surveys pre- and post-nourishment were used to evaluate the response of Cempedak Bay beach to the beach nourishment programme. Several analyses have been carried out including the distribution pattern of sand volume changes and beach level changes, development of the beach width as well as the application of the parabolic bay model. The Mepbay model was used (Klein et al., 2003) to classify the planform stability of Cempedak Bay beach. Derived from the parabolic bay shape equation (PBSE), this model was applied to identify three points that are important in diffractions (i.e. the updrift control point, the downdrift control point and the end point along the downdrift tangent on the beach) and the final output is the static equilibrium shoreline. The planform stability is then determined based on criteria suggested by Hsu et al., (2010). The criterion is defined in **Section 2.2** and **Figure 5-1** shows examples of static, dynamic and natural reshaping of the bay, which explicitly describes the identification of planform stability.

5.4 Results of field survey analyses

The analyses of the Cempedak Bay beach response to beach nourishment are grouped into four main categories: distribution pattern of sand volume changes, distribution pattern of beach level changes, the development of beach width, and application of the parabolic bay model to determine the stability of Cempedak Bay beach.

5.4.1 Distribution pattern of sand volume changes

In terms of engineering application, the total loss of sand over time is the main point of concern. To gain insight into what is happening in the system, graphs of the total volume of sand and the percentage of sand loss and gain within the nourishment region compared to the preceding period are displayed in **Figure 5-7 (a) and (b),** respectively. The volume is computed based on topographic and bathymetry survey data over a period of monitoring years. The shoreline profiles were spaced at an alongshore interval of 25 m and an interval of 10 m along the profile. The landward limit of the shoreline profiles was 100 m beyond the high water mark (+ 0.95 m LSD) and the seaward limit was 1000 m seawards of the high water mark. It should be noted that 177 000 m^3 of sand was placed on the beach and the foreshore area in July 2004. The new beach material was a mixture, composed of coarse and fine sand. The latter was placed on the lower part of the beach (below mean sea level) and coarser sand was placed on the upper part of the beach (above mean sea level).

In total, a net volume of 47 000 m^3 or 27 % of sand decreased from the nourishment zone over the monitoring period (July 2004 to July 2007), but the majority of this was lost in the first 3 months. Three months after nourishment (August 2004 to October 2004), 21 % or 37 000 m^3 of sand in the system was lost due to stabilising effects. Following the initial loss, only a further

10 000 m³ (6%) was lost over the next 2.5 years. As a result, the rates of sand losses and sand gains were comparable with each other with slight losses and gains during the southwest and north-east monsoon times, respectively, as explained in **Figure 5-7**. This pattern can be seen for the whole year of 2005, and most likely a similar trend was repeated for 2006 and 2007 as also seen in **Figure 5-7(b)**. The percentages of loss during the south-west monsoon were approximately 3 to 7%, whereas during the north-east monsoon the percentage of sand gain was slightly higher at around 6 to 9 %. This trend implies that seasonal climate modifies the progression of sand volume over the monitoring period. The temporal distribution of sand volume pattern [**Figure 5-7(a)**] infers that the nourished beach has remained stable after the stabilising period of 3 to 4 months and normal fluctuation occurs over a period of 33 months. This consequently contributes to small percentages of sand gain and loss which were approximately 9 to 27%, respectively [**Figure 5-7(b)**].

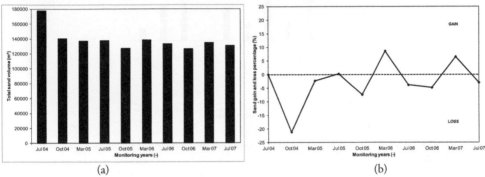

(a)   (b)

Figure 5-7. Total sand volume (m³) (a) and sand loss and gain percentage (b) within the nourishment area compared to the preceding period.

Further graphs of the sand gain and loss at three different locations outside the nourishment zone are shown in **Figure 5-8** in order to show the probable sand movement into these regions. Immediately after the nourishment, 70 000 m³ had moved into the centre offshore area, while 45 000 and 13 000 m³ had been naturally distributed into the north offshore and south offshore area, respectively. However, after the stabilisation period, the net amount at the southern area over the total monitoring period showed a greater volume of sand with a net volume of 65 000 m³ in comparison with the other two zones. The centre offshore area received less volume of sand (24 000 m³) whereas the northern area indicated a slight loss of sand (4000 m³). The analysis shows that the net sand volumes for the southern area and centre offshore zone were positive which implies the movement of sand into these regions. The northern area experienced a significant loss of sand, instead.

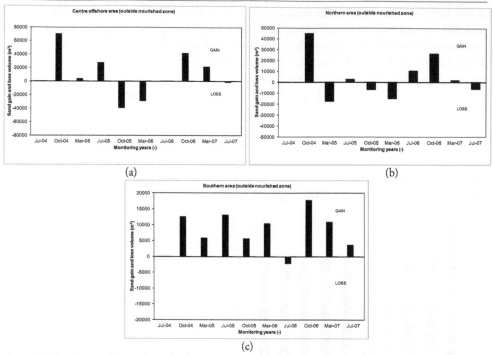

Figure 5-8. Sand gain and loss volume (m3) at three zones outside the nourishment region: (a) Centre offshore area; (b) northern area; (c) southern area

Figure 5-9 presents the spatial distribution of beach volume changes for the pre-nourishment, nourishment, and post nourishment phases. For this analysis, the total volume per metre is determined based on the average beach level from the seawall towards the 120 m extending offshore beyond the 0 m coastline as in **Figure 5-9**. This is to ensure that the full beach profile (and therefore volume) can be captured, especially after the re-nourishment period. It should be noted that the nourishment period was from May to July 2004 and October 2004 was the first month of the monitoring programme after the nourishment. The general pattern of sand distribution across the entire nourishment profiles shows that the northern profiles experienced a higher volume of sand than the southern profiles. Indeed, a greater loss of sand was recorded at profile 1300, while the middle stretch of the beach showed small variations of sand volume compared to the preceding period. The northern profiles, especially at profile 400 and profile 500, had gained more sand 3 months after the nourishment period than in the nourishment year. Moreover, the October 2005 profile and October 2006 profile showed similar behaviour of sand accretion.

Figure 5-10 shows the spatial distribution of sand gain and sand loss in comparison with the preceding period. Ignoring sand volume profiles at nourishment time, the changes in sand volume are obviously seen at the northern profiles and southern profiles. The middle stretch profiles show small variations within the range 50 to 280 m³/m which indicates a normal fluctuation over the post-nourishment period. It is likely that the northern profiles and southern profiles experienced significant sand gains and sand losses or vice versa. It appears that this may happen due to the alternating direction of sediment transport for each season.

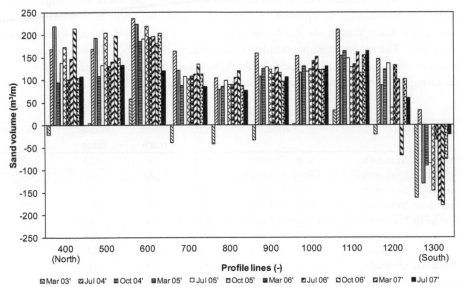

Figure 5-9. Spatial distribution of sand volume for pre-nourishment, nourishment, and post-nourishment.

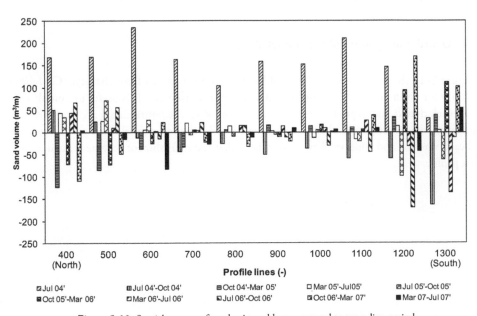

Figure 5-10. Spatial pattern of sand gain and loss compared to preceding period.

In order to prove this hypothesis, the zone was divided into three parts, namely the north reach, middle reach and south reach. The tabulated net volume of sand presented in **Table 5-3** describes the distribution pattern of sand towards the north and south profiles, which eventually supports the above hypothesis. The north reach gains sand during the south-west monsoon while the north-east monsoon proves otherwise. On the other hand, the south reach gains more sand than the north reach during the north-east monsoon and loses sand during the south-west monsoon time. The net sand volume calculated from March to July 2007 shows a reverse pattern of common trend found in the present analysis; however, this could be neglected since the volume changes at the north reach and south reach are small compared to the other period.

Table 5-3. Net sand volume at three different reaches

Season	Period	Year	North reach	Middle reach	South reach
Southwest Monsoon	July - October	2004	74	-223	-220
Northeast Monsoon	October - March	2005	-208	-24	73
Southwest Monsoon	March - July	2005	68	14	19
Southwest Monsoon	July - October	2005	105	-6	-158
Northeast Monsoon	October - March	2006	-144	-5	207
Southwest Monsoon	March - July	2006	53	71	-167
Southwest Monsoon	July - October	2006	123	-62	-179
Northeast Monsoon	October - March	2007	-156	-11	273
Southwest Monsoon	March - July	2007	-12	-95	-12

* +ve / -ve value indicates accretion and erosion, respectively. North reach=profile 400 to profile 500; Middle reach=profile 600 to profile 1100; South reach=profile 1200 to profile 1300.

5.4.2 Distribution pattern of beach level changes

Figure 5-11 shows the cross-section profiles at the nourished area from chainage (CH) 400 to CH 1300 for pre-nourishment, nourishment and post-nourishment. It is useful to note that July 2004 (black dotted line) represents the baseline period when the nourishment activity took place. Apparently, the upper part of the beach is convex unlike the pre-nourishment profiles in which the beach was low and concave. Thus it is evident that the nourishment scheme created a higher beach profile.

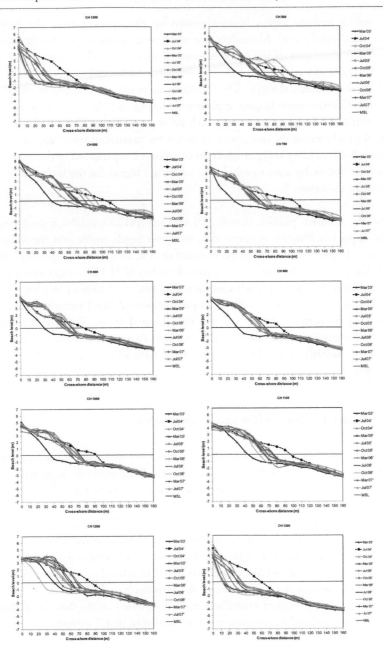

Figure 5-11. Cross-section profiles at the nourished profile zone from CH 400 until CH 1300. MSL refers to mean sea level (+0.05 LSD) in m.

Looking at profiles CH 400 to CH 800, there is a 'nodal point' present at around 40 to 50 m from the profile head, which separates the two different profile patterns. The north-east monsoon profiles indicate higher beach elevation on the upper part of the beach which is landward of this 'nodal point'. Meanwhile, the south-west monsoon profiles show greater beach elevation on the lower part of the beach seaward of this point. The steepening of these beach profiles is the result of a reshaping of the nourishment. Additionally, this profile behaviour was influenced by the seasonal changes. The variations in beach crest at the northern profiles are linked to the different wave heights experienced during the different seasons. For instance, during the north-east monsoon the waves reach up to 4 m (**Figure 5-6**), building higher beach crests up to 4 m whereas in the south-west monsoon the maximum waves appear to be around 2 m (see **Figure 5-6**), yielding lower beach crests of around 2 m, seaward of the 'nodal point'. The beach crest is linked to the wave run-up levels, which are higher for the larger waves experienced in the north-east monsoon.

Likewise, the southern profiles, namely CH 900 to CH1200, exhibit a similar beach profile pattern; the upper beach profile is steeper than the lower beach profile over the whole year cycle. However, there is not a seasonal variation in beach crest as seen in the northern profiles. On the other hand, the profile at the southernmost end of the bay (CH 1300) experiences a great loss of sand thus significantly lowering beach elevation. The drastic change in bed level at this profile could possibly be caused by wave diffraction process around the Tembeling headland and the presence of megarip currents. In fact, the profile is located at the convex rocky headland itself and it is thus difficult to hold a beach at such location. As the scale of the headland is relatively small, sand can easily bypass the Tembeling headland with a lowering of the beach levels at CH 1300 as a result.

5.4.3 Development of beach width and beach rotation

The development of beach width is one of parameters that can be used to determine the response of eroded beaches to sand nourishment. **Figure 5-12** shows the temporal distribution of an average dry beach and the spatial development of average dry beach width before, during and after nourishment. The dry beach width is defined as the distance from the 0m depth contour landward up to the seawall boundary. The evaluation of dry beach width provides information about recreational and habitat space availability and morphological changes resulting from maximum wave run-up (Roberts et al., 2010). In terms of temporal distribution, it can be observed that the dry beach width increased 60 to 70 m after the 1-year nourishment exercise compared to prenourishment (average 40 m). A sharp decrease in October 2004 indicates that Cempedak Bay beach experienced a stabilisation process. Based on the data set of March 2005 to July 2007, the present shoreline recession rate is estimated to be 1.7 m/year.

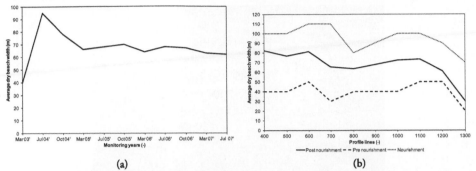

(a)   (b)

Figure 5-12. Beach width variation of the Cempedak Bay beach. (a) Temporal pattern of beah width and (b) spatial distribution of beach width at three different phases.

The spatial development of beach width shows clear feedback within the nourishment profiles. The nourishment profile as in **Figure 5-12(b)** is plotted based on the profile surveys at the end of the nourishment period, which was July 2004. The solid black line, which indicates the average dry beach width limit after the nourishment exercise has been calculated, based on the averaged beach width for each seasonal period recorded for 3 years after the placement of the material. Obviously, the width of the beach before the restoration programme was lower at an average of 40 m. However, the beach width recedes after 3 years and coincidently follows the similar pattern as per pre-nourishment. The northern profiles experience a wider dry beach area and the southern profiles show lower beach width, especially at profile 1300. The middle stretch of the nourished profiles seems to have constant width of around 70 m. To gain more insight into the shoreline development of Cempedak Bay beach, the 0m contour line is geo-referenced, thus mapping the contour lines on the satellite imagery. **Figure 5-13** shows the shoreline development over the 3-year monitoring survey. It is clearly seen that the post-nourishment shoreline slightly recedes; however, it is still greater that the pre-nourishment shoreline.

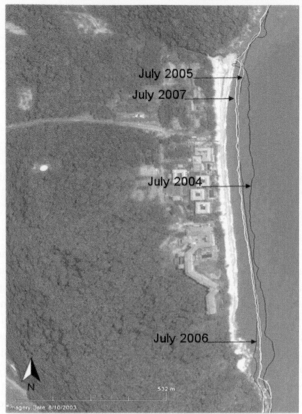

Figure 5-13. Shoreline development of Cempedak Bay beach. Background image represents beach planform 2003. (Satellite image is adapted from Google Earth version 6; image © 2012 DigitalGlobe)

Figure 5-14 shows the shoreline pattern of three beach plan shapes (July 2006, March 2007 and July 2007), which is overlaid on the satellite image of 2006. The idea is to observe whether or not the beach has undergone a rotation process. Apparently, a slight rotation does exist where a hinge point is located somewhere further downcoast of the southern embayment. Looking at the close-up photos on the right side of **Figure 5-14**, image A (northern side) shows the recession of the shoreline in March 2007 compared to July 2006. On the other hand, the shoreline in March 2007 (image B, southern side) tends to move a bit seaward. This finding coincides with the sand volume pattern (see **Table 5-3**), which infers that seasonal wave climate influences the sediment transport pattern thus leading to a small rotation process. Despite the fact that the beach near the southern headland regained sand in July 2007 in comparison with July 2006, analysis shows that the sand volume (**Figure 5-7(b)**) as well as bed profile level (**Figure 8**) at this particular area is significantly low and the beach area is severely eroded.

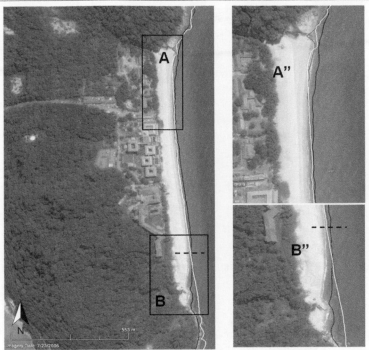

Figure 5-14. Beach width development and beach planform rotation. The 0 m contour line for black line indicates shoreline planform of July 2006 , white line is March 2007 and grey line is July 2007. The satellite image shows beach condition in July 2006. Two photos (A" and B") on the right side are the close-up images of image A and B, respectively. A dotted line signs a hinge point where the beach starts to rotate. (Satellite image is adapted from Google Earth version 6; image © 2012 DigitalGlobe)

5.4.4 Application of the parabolic bay model

The computed shorelines and beach stability of Cempedak Bay beaches are presented in this section. Based on the offshore wave roses plotted in **Figure 5-6**, the predominant wave direction is from north/north-east for the north-east monsoon and south-west/south/south-east for the south-west monsoon. Due to limitations of the parabolic bay model, which assumes that the planform shape of the bay is caused by one predominant wave direction, it was assumed that the north-east monsoon is the most significant. This was because the dominant wave direction based on the SSMO wave data is from the north-east monsoon season. The highest wind speed and wave height are associated with this sector. Based on satellite images and existing beach orientation (**Figure 5-15**), the shadow zone is located in the northern part where the wave diffraction point is located. As the upcoast headland is dominated by the rocky area, and due to the unknown position of the wave diffraction point, two virtual points of headland tip were chosen. One is located above the water line while another one is placed further offshore from the first virtual point. As an input to parabolic bay model, a 2007 satellite image was selected.

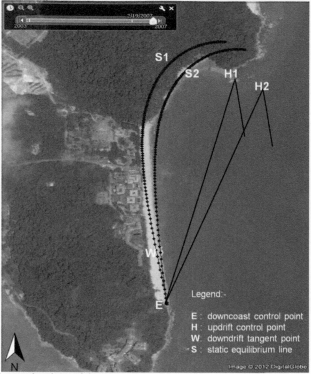

Figure 5-15. Predicted static shorelines of the Cempedak bay beach. (Satellite image is adapted from Google Earth version 6; image © 2012 DigitalGlobe)

Figure 5-15 shows the result of predicted shoreline at Cempedak Bay. In the application of the parabolic bay model, the approaching wave should be parallel to the downdrift shoreline. Due to the lack of information regarding the actual diffraction point around the upcoast headland, therefore, two virtual diffraction points of headlands, namely H1 and H2, were proposed for the study area. H1 represents the point located near the land (above the waterline) whereas point H2 is placed further seaward below the water line. It is expected that the tip of the headland may be located submerged under the water due to the geological setting. Additionally, as the headland is surrounded by submerged rocks, it is also possible that the control point might probably be situated offshore. Tangential point (point W) and downcoast control point E both remain at the same position. By applying the upcoast control point (H1), it is found that the static line (S1) is slightly shifted landward. However, the static line tends to move seawards in relation to the present coastline when the tip of the headland is located below the water line. It is hard to conclude in which category the beach should be classified. As the model prediction is subjective because of the selection of the diffraction points, it was not possible to draw any conclusion with regard to the stability of Cempedak Bay beach. Further discussion on the suitability of the model in predicting static equilibrium shoreline is presented in **Section 5.4**.

5.5 Discussions

A preliminary study of Cempedak Bay beach was performed. A sediment budget analysis for pre-nourishment, nourishment, and post-nourishment periods was studied at Cempedak Bay beach and the parabolic bay model applied to inspect the stability of an eroded beach.

5.5.1 Distribution pattern of sand volume changes

The result of the temporal pattern of total sand volume changes for the post-nourishment period shows that Cempedak Bay beach lost about 27 % or 47 000 m³ volume of sand (net loss), which is equivalent to 16 m³/m per year from the nourishment zone over the 3-year monitoring period. The first 3 months after the nourishment witnessed a great loss of sand, namely 21 % or 37 000 m³ of sand. This was due to the packing process and stabilising period. Beyond this period, a normal fluctuation of sand gain and sand loss occurred with the net sand loss of approximately 10 000 m³ (4.00 m³/m per year). The rates of sand losses and sand gains were comparable with each other within the range 9 to 27%. The overall losses of sand were greater during the south-west monsoon than during the north-east monsoon. The evaluation of the sand volume evolution outside the nourishment zone reveals the movement of sand into these regions. This contribution may stem from the sand nourishment scheme or possibly from sediment that bypassed the headlands. As indicated in **Figure 5-8**, after the stabilising period, the southern offshore area received more sand, approximately 65 000 m³ (net) compared to the centre offshore zone (24 000 m³). On the other hand, the northern offshore zone experienced a slight loss of sand (4000 m³).

The net sand gain and sand loss analysis shows that Cempedak Bay beach has undergone two different processes, especially at the northern part and southern part of the beach which experience significant variation over the post-nourishment periods (see **Table 5-3** for comparison). This variability may be the result of the monsoon-related hydrodynamic variability. During the south-west monsoon, the sand losses over the 3-year period after the beach nourishment show that the northern part of the beach features a higher sand volume than the southern part (refer to **Figure 5-10** and **Table 5-3** separately). However, this result reverses for the north-east monsoon. This finding is supported by the fact that the sediment is mainly transported from the northern to the southern region during the north-east monsoon period and vice versa during the south-west monsoon period (DID, 2005). From this evidence, it is concluded that the alternating sediment volume patterns in Cempedak Bay beach are caused by seasonal rhythmic behaviour; however, this does not significantly influence the shoreline behaviour. Some other researchers have found similar findings to those of the present study. For instance, Norcross et al., (2002) reported the rhythmic behaviour of seasonal transport patterns at Kailua beach given the seasonal wind and wave climate. They then concluded that the alternating erosion and accretion are the result of longshore transport. Morton et al., (1995) found that alternations in volume increases and decreases at adjacent sites are due to wave-driven cross-shore transport, whereas longshore transport explains systematic increases or decreases in sand volume. However, this is contrary to the present deduction which suggests that the alternating erosion and accretion patterns are a result of longshore transport. Clarke and Eliot (1988) concluded that the short-term changes in beach morphology resulted from longshore transport, whereas long-term changes could be explained by onshore–offshore sediment movement. Their findings are in line with Norcross et al., (2002) who indicated that longshore transport dominates the annual to interannual shoreline variability whereas decadal variability is dependent on cross-shore profile changes governed by sediment availability.

Figure 5-16 explains the sediment mobility for the two different seasonal systems within Cempedak Bay beach. According to this figure, the north-eastern waves drive the longshore sediment to the southern region where the accumulation of sand is expected to occur. Likewise, given the direction of seasonal wind and wave climate during the south-west monsoon period, it would seem likely for sand to be transported from the southern end of the bay towards the northern end of the bay. As a result, the northern area experiences higher beach volume than the southern area during that period. **Table 5-4** provides a picture with regard to alternating erosion and accretion patterns with respect to seasonal changes. It should be noted that the middle reach which encompasses profile 600 to profile 1100 is excluded from this table as those profiles do not vary significantly. Readers are suggested to refer to this table together with **Table 5-3** above for better explanation.

Table 5-4. Sand volume behaviour with respect to seasonal variations

Season	Period	Year	North reach	South reach
Southwest monsoon	July - October	2004	Accretion	Erosion
Northeast monsoon	October - March	2005	Erosion	Accretion
Southwest monsoon	March - July	2005	Accretion	Accretion
Southwest monsoon	July - October	2005	Accretion	Erosion
Northeast monsoon	October - March	2006	Erosion	Accretion
Southwest monsoon	March - July	2006	Accretion	Erosion
Southwest monsoon	July - October	2006	Accretion	Erosion
Northeast monsoon	October - March	2007	Erosion	Accretion

* North reach=profile 400 to profile 500; South reach=profile 1200 to profile 1300.

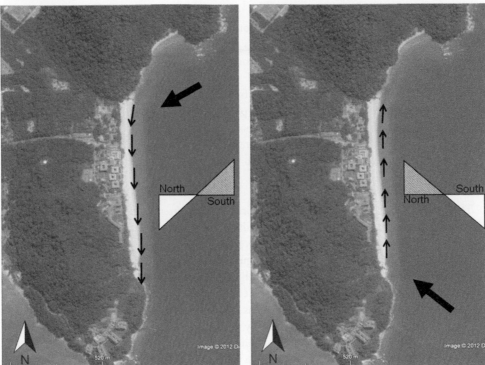

Figure 5-16. The mobility of sediment for different seasonal systems. Left figure is the north-east monsoon, and right figure is the south-west monsoon. Black small arrows indicate the direction of sediment transport, and black bold arrow represents the direction of wave for the respected seasons. Insets show the accretion and erosion of sand. A downward diagonal pattern in the triangle indicates the accretion of sand and erosion signs in a white triangle. (Satellite images are adapted from Google Earth, 2012)

5.5.2 Distribution pattern of beach level changes

The beach morphology of Cempedak Bay beach depicts a typical convex profile (**Figure 5-11**) after the nourishment period, indicating the depositional activities prevailing in the coastal zone in almost every season. The wave climate and coastal configuration is such that the beach experiences deposition especially during the north-east monsoon, and the cross-shore sediment transport is anticipated predominantly towards the onshore. In addition, due to its bay-like configuration in which sediments are trapped by the headland located on either side of this bay, the accumulation of sand is possibly caused by the longshore drift. The presence of a nodal point, which separates between two different profile patterns within profile 400 to profile 800, marks the influence of the monsoonal systems. The north-east monsoon profiles exhibit higher beach profile on the upper part of the beach whereas the south-west monsoon profiles show a greater beach elevation on the lower part of the beach seawardly beyond this turning point. Therefore, the steepening of the beach profiles concludes that this variation is influenced by the seasonal change and leads to reshaping of the sand renourishment. Additionally, the variations in beach crests are linked to different seasonal climates (**Figure 5-5**) as explained in **Section 5.3**. Wave heights up to 4 m in the north-east monsoon will result in higher runup levels and therefore higher beach crest than in the southwest monsoon, which has smaller wave height at around 2 m. **Figure 5-11** at profile 400, for instance, shows a crest height of around 4 m following the north-east monsoon and around 2 m following the south-west monsoon. In addition, the changes in beach profile configuration (steep slope and gradual slope denote erosion and accretion, respectively) might be related to the onshore–offshore transport by waves. This onshore–offshore movement of sand was reflected in beach morphology by the construction and destruction of sub-aerial berm growth and migration of nearshore sand bars in the surf zone. The positive sand volume (24 000 m³) after recharge at the centre offshore area as shown in **Figure 5-8** may be evidence of this cross-shore sand transport which thus enhances the mobilisation of sand into this zone.

The southern part of the beach showed a typical profiles pattern in line with northern profiles. The upper part of the beach is steeper and the lower part is flatter. This, however, is not influenced by the seasonal variation in the beach crest as seen in the northern profiles. In particular, profile 1300 is extremely eroded (**Figure 5-11**). This result may be due to the steeper beach profile at the southern part of the bay which indicates that this part is more reflective to the waves, thus leading to the erosion process. It is useful to note here that this profile is located on the convex rocky headland (**Figure 5-3**), thus inferring that the beach could not possibly hold on to it. One of the possibly actions that may result in rapid beach losses is the presence of megarip currents which can carry huge quantities of sediment offshore (Short, 1985). More recently, Short (1996) discussed the impact of a headland on the embayment. As the headlands are closer together and the wave height increases, the entire bay circulation may eventually become impacted by the headlands. At this stage, topographically controlled cellular megarips, prevail. The drastic change at this profile coincides with the volume of sand loss as clearly indicated in **Figure 5-9**. Moreover, the southern headland (Tanjung Tembeling) may not be large enough to act as a complete sediment trap and the reduction in beach levels and volume at profile 1300 may be due to a loss of sediment southwards. This finding is supported by **Figure 5-8(c)** where 65 000 m³ volume of total sand has been naturally deposited into the southern region outside the nourishment zone over the monitoring period. Further analysis should be carried out to investigate the possible sediment loss at this particular location.

5.5.3 Development of beach width and beach rotation

The results concerning the average dry beach width show that Cempedak Bay beach was wider after nourishment compared to the pre-nourishment period. The temporal pattern of average dry beach width as in **Figure 5-12(a)** shows that the beach width increased from 60 to 70 m after the 1-year nourishment exercise and remained constant throughout the next monitoring period compared to pre-nourishment. The present shoreline recession rate was estimated to be 1.7 m/ year for the data set of March 2005 to July 2007, relatively higher than the shoreline recession rate reported in the NCES, which was 0.8 m/year (Stanley Consultants, 1985). The development of beach width provides a wider dry beach area, thus enhancing recreational activities. However, the temporal distribution of beach width indicates that the beach widths slightly reduced after 3 years' monitoring. The post-nourishment beach width follows a similar pattern as the pre-nourishment one. The northern profiles feature higher beach width than the southern profiles which coincides with the evaluation of sand volume changes and beach level changes as presented in **Figures 5-9 and 5-11**, respectively. The variation of beach widths at the northern profiles and southern profile was further observed to determine the influence of the beach rotation process. Beach rotation is a natural phenomenon that occurs on beaches with extremities limited by headlands, as well as on coastlines where there are rigid structures. In most cases, rotation is due to the seasonal or periodic wave regime and in particular to the direction of the waves (Klein et al., 2002; Ranasinghe et al., 2004; Short and Trembanis, 2004; Short et al., 1995: 2000). Referring to **Figure 5-14**, it appears that the shoreline planform of March 2007 at the northern part of the beach shows a recession line compared to the shoreline planform of July 2006, whereas the southern part of the beach proves otherwise. This infers that Cempedak Bay beach experiences a small rotation process where the hinge point is located further away downdrift of the beach. The rotation process does exist due to the different seasonal wave climate. This is in agreement with the sediment movement pattern as discussed in **Section 5.1**, above.

As the headland is small and the longshore drift is relatively high, sand can partially bypass (Bray et al., 1995; Short, 1999) and continuously be lost from the bay. If this is the case, a small beach rotation might be expected to occur or be completely absent. However, if the headland is big, all sediment will be blocked and remain inside the bay (Bray et al., 1995); beach rotation may occur considering the similar pattern of different wave climate as discussed herein. It appears that Cempedak Bay is showing classic signs of beach evolution within a (semi-) enclosed pocket bay; that is, fluctuations in beach positions (and volumes) at either end with a hinge point somewhere in the middle (which shows little change in beach positions or volume). There are some net losses and this may be because the headlands in this bay are reasonably small and do not hold on to all of the material. There will also be some losses offshore during large storm events.

Figure 5-17 presents the beach development for an associated schematised beach for pre-nourishment and postnourishment periods. The pre- and post-nourishment period indicates the beach area is separated into two distinct zones. The post-nourishment diagram and associated photos show the beach tends to create a higher profile a few metres landward of the shoreline, which indicates the accumulation of sand (berm formation) due to onshore transport processes.

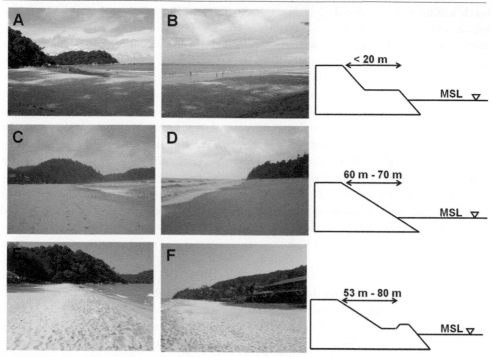

Figure 5-17. Morphology of the Cempedak bay beach. Left photos are view from north and middle photos are south views. (A and B) pre-nourishment (2003), (C and D) post-nourishment (2005), and (E and F) post-nourishment (2006). The diagrams on the right side represent the schematized beach width at the Cempedak bay which indicate the distinction of two separated zones.

5.5.4 Application of the parabolic bay model

Based on the results of the PBSE model, there are some uncertainties, which affect the conclusion to categorise Cempedak Bay beach. Multiple wave diffraction points have been proposed at the upcoast headland for Cempedak Bay beach as displayed in **Figure 5-15**. The selection of this point is very subjective due to the geological setting of the intervening headlands. As the upcoast headland is composed of a rocky area (**Figure 5-18**), difficulty was found in setting the actual location of the diffraction point. The exact point might be submerged under water and further away from the headland. Jackson and Cooper (2010) present some observations regarding the practical application of the equilibrium planform concept and one of them is the subjectivity in selecting the diffraction point. This aspect has also been addressed by Lausman et al., (2010) who demonstrated through blind testing that the operator (respondent) is strongly influenced by the sight of the actual shoreline position. Therefore, it is crucial to have information regarding the geological condition of the headland as it is important to locate the exact point of wave diffraction. As a result, an accurate static shoreline would be predicted, and a reliable beach state could be achieved. On the other hand, the change of the tide, wave direction and wave height probably influence the selection of this point.

Figure 5-18. The Tanjung Pelindung headland consists of boulders and rocks.

The selection of the downcoast control point and downcoast tangent point is also questionable. The parabolic bay model assumes that the approaching wave is always parallel to the downcoast coastline. However, this is not always the case. For example, if the predominant wave is specifically propagating 55° from the north (not presented in this study), the predicted static shoreline significantly strayed away from the present coastline.

The other possibility that may influence the stability of Cempedak Bay beach is the discharge from the Cempedak river which is located at the northern part of the beach, as presented in **Figure 5-19**. The river has only a minor catchment area (1.2 km²) and the discharges only influence the beach where the water runs across it. The sand transport in the river is assumed to be very low and with no influence on beach morphology. The river discharge is estimated to be 0.1 m³/s in the south-west monsoon and the sediment concentration is defined as 1 mg/l (DID, 2007). Although the sediment transport rate from the river is low (DID, 2007), this however could influence the morphology of the northern part of Cempedak Bay beach. Jackson and Cooper (2010) discuss the impact of the River Dall to the stability of the Cushendall embayment. They found that the overall coastline prediction is good with the exception of the southern section where the river emerges and the actual shoreline is located seaward of the one predicted by the model. Additionally, van Rijn (1998) emphasises two factors that influence the curvature shape of embayed beaches, which are the sand bypassing from the updrift headland and sediment input of a river outlet inside the bay. Moreover, Silveira et al., (2010) conclude that coasts with greater fluvial sediment input tend to have a larger number of beaches in the dynamics state. Further studies are required to inspect the volume of sediment discharged into this pocket bay beach.

Figure 5-19. The Cempedak river downstream outlet drains water to Cempedak bay beach.

As the model applied in this study is only capable of empirically predicting the static equilibrium line, the physical processes along the beach and around the headland are still unknown. In fact, the parabolic bay shape equation does notdescribe the time frame in which the bay would reach the static shape; it only gives the final result of the predicted shoreline (Lausman et al., 2010). It would be beneficial if the PBSE model could be used together with the process-based model to further examine the stability of the beach as well as the detailed hydrodynamic and morphological processes. Although Mepbay is a simple empirical model, it does not provide any information with regard to timescales for a bay to reach equilibrium, but it is still a useful predictive tool for preliminarily assessing beach stability and headland control as long as enough detailed information about the specific study area is available.

5.5.5 Sand bypassing across Tanjung Tembeling headland

Based on the analyses presented in this study, it is inferred that sand bypassing may possibly across the southern headland i.e Tanjung Tembeling headland. Much of the nourished sand is distributed to the southern part of Cempedak bay. The occurrence of beach rotation proves that the beach area at the southern part is wider than the northern part of the beach. Sand at the southern area is easily bypassed Tanjung Tembeling headland as the headland is reasonably small. **Figure 5-20** clearly shows the dispersion of sand wave at the lee side of Tanjung Tembeling headland as a result of the subaqueous sand moving around the headland. There is a possibility of a rip current to occur at the toe of Tanjung Tembeling headland as the significant loss of sand had been observed at profile CH 1300.

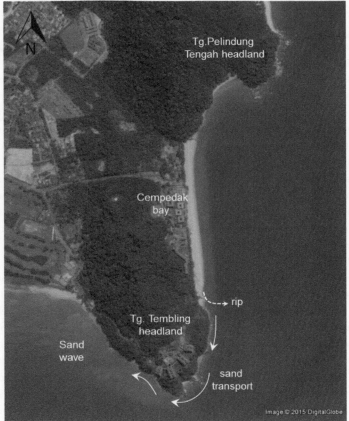

Figure 5-20. Sand bypassing across Tanjung Tembeling headland. Image courtesy of Google Earth version 7; image ©2015 DigitalGlobe.

5.6 Conclusions

The response of Cempedak Bay beach to beach nourishment has been investigated. A sediment budget analysis based on beach profile surveys reveals the variability along the profile transects. A planform stability indicator recommended by Hsu et al., (2008) to classify the stability of headland bay beaches was unsuccessful in categorising the state of stability of Cempedak Bay beach. Based on the results presented in this study, several detailed conclusions can be drawn.

1. The analysis of temporal sand volume patterns showed that Cempedak Bay beach has lost about 27 % or a total of 47 000 m³ volume of sand which is equivalent to 16 m³/m per year from the nourishment zone over the 3-year monitoring period. The first 3 months after nourishment witnessed the greatest loss of sand with the percentage of 21% or 37 000 m³ of

sand. In general, the overall loss of volume excluding the initial rapid loss of sand after recharge was 10 000 m³, equivalent to 6 % over the next 2.5 years. Normal fluctuation of temporal sand gain and sand loss illustrates the stable condition of Cempedak Bay beach over the monitoring periods.

2. The analysis of seasonal changes assessed through temporal beach volume patterns, which indicate the shoreline variability, can be characterised by an alongshore rhythmic pattern of alternating seasonal behaviour. The north-east monsoon drives the sediment towards the southern part of the beach, thus featuring higher beach volume than the northern region. However, this proves otherwise for the south-west monsoon. A simple seasonal transport pattern has been proposed to account for alternating erosion and accretion.

3. The temporal distribution pattern of beach level changes reveals the existence of a nodal point, which is influenced by the monsoonal system. The north-east monsoon profiles exhibit greater beach elevation on the upper part of the beach whereas the south-west monsoon profiles feature higher beach elevation on the lower part in the seaward direction beyond the nodal point. This pattern was linked with the seasonal variation in both wave direction and heights. Although the southern profiles showed a similar profile pattern to the northern profiles, there was no seasonal variation in the beach crest seen in the northern profiles.

4. The most southern profile (profile 1300) experiences significant drop in beach level as well as great sand losses. Sand could easily bypass the headland as the headland itself is relatively small. In fact, the profile is positioned on the convex rocky profile and it is difficult to hold a beach at such a particular location. As Cempedak Bay is small and impacted by the headlands, there is a possibility that megarip currents exist thus transporting a huge amount of sediment away from the beach. The sand volume analysis at the southern area outside the nourishment zone positively shows the contribution of sand into this region and thus verifies the potential growth of sand southwards around this headland. A long groyne may be helpful to trap sediment from bypassing the southern headland. However, careful assessment should be carried out to investigate the potential consequences from this beach control measure.

5. The analysis of dry beach width shows the beach area increased 60 to 70 m after the nourishment and remained stable over the monitoring years. The present shoreline recession rate is estimated to be 1.7 m/year for the data set of March 2005 to July 2007. The spatial distribution of the beach width indicates the northern beach area is wider whereas the southern beach area experienced lower beach width, which was coincident with the temporal pattern of sand volume and beach profile changes. A small beach rotation does exist and is attributed to a seasonal or periodic shift in wave climate, in particular wave direction.

6. It appears that Cempedak Bay is showing classic signs of beach evolution within a (semi-) enclosed pocket bay, namely fluctuations in beach positions (and volumes) at either end with a hinge point somewhere in the middle (which shows little change in beach positions or volume). There are some net losses and this may be because the headlands in this bay are reasonably small and do not hold onto all of the material. There will also be some losses offshore during large storm events.

7. The planform stability of Cempedak Bay beach is not easy to determine due to model uncertainties and especially the selection of the diffraction point. This should be performed carefully as it may change with tide, with wave direction and with wave height. Site investigation on the rocky headland of Pelindung Tengah should be carried out as the

bathymetry around the rocky headland might be different. The tip of the headland might be located submerged under the water and offshore in order to confirm the static stability of the beach. The use of a process-based numerical model is needed to simulate the morphodynamics of Cempedak Bay beach. The headland sediment bypassing could possibly be assessed, thereby revealing the potential loss of southern beach area.

8. As the empirical model can only predict the future shoreline without considering the time frame for the development of the headland bay beach, a coastline model or process-based model which takes into account the morphological change processes is highly recommended. As a result, the development of bay beaches could be hindcasted, and both accretion and erosion patterns could be established, from which informed inferences can be modelled for future management activities.

References

Bray, M.J., Carter, D.J., and Hooke, J.M. (1995). Littoral cell definition and budgets for central southern England. *Journal of Coastal Research* 11(2): 381–400.

Clarke, D.J., and Eliot, I.G. (1998). Low-frequency changes of sediment volume on the beachface at Warilla Beach, New South Wales, 1975–1985. *Marine Geology* 79(3–4): 189–211.

Cooper, N.J., Legett, D.J., and Lowe, J.P. (2000). Beach profile measurement, theory and analysis: practical guidance and applied case studies. *Journal of Chartered Institution of Water and Environmental Management* 14(2): 79–88.

Department of Irrigation and Drainage Malaysia (DID), (2005). Annual Report 2005. DID, Kuala Lumpur, Malaysia.

Department of Irrigation and Drainage Malaysia (DID), (2007). Detailed Design Report of Projek Perintis Pemuliharaan Pantai Pelancongan Menggunakan Pressure Equalization Module (PEM) di Teluk Cempedak, Kuantan, Pahang. Coastal Engineering Division, Kuala Lumpur, Malaysia.

Gares, P.A., Wang, Y., and White, S.A. (2006). Using LIDAR to monitor a beach nourishment project at Wrightville Beach, North Carolina, USA. *Journal of Coastal Research* 22(5):1206–1219.

Ghazali, N.H. (2006). Coastal erosion and reclamation in Malaysia. *Aquatic Ecosystem* Health and Management 9(2): 237–247.

Gonzales, M., Medina, R., Gonzales-Ondina, A. et al. (2007). An integrated coastal modeling system for analyzing beach processes and beach restoration projects, SMC. *Computers& Geosciences* 33(7): 916–931.

Hsu, J.R.C., and Evans, C. (1989). Parabolic bay shapes and applications. *Proceedings of the Institution of Civil Engineers*, Part 2 87(4): 557–570.

Hsu, J.R.C., Benedet, L., Klein, A.H.F. et al. (2008). Appreciation of static bay beach concept for coastal management and protection. *Journal of Coastal Research* 24(1): 812–835.

Hsu, J.R.C., Yu, M.J., Lee, F.C., and Benedet, L. (2010). Static bay beach concept for scientists and engineers: a review. *Coastal Engineering* 57(2): 76–91.

Jackson, D.W.T., and Cooper, J.A.G. (2010). Application of the equilibrium planform concept to natural beaches in Northern Ireland. *Coastal Engineering* 57(2): 112–123.

Klein, A.H.F., Benedet, L. and Schumacher, D.H. (2002). Short term beach rotation processes in distinct headland bay beach system. *Journal of Coastal Research* 18(3): 442–458.

Klein, A.H.F., Vargas, A., Raabe, A.L.A. and Hsu, J.R.C. (2003). Visual assessment of bayed beach stability with computer software. *Journal of Computer & Geoscience* 29(10): 1249–1257.

Komar, P.D. (1983). Beach processes and erosion. In: CRC Handbook of Coastal Processes and Erosion (Komar PD (ed.)). CRC Press, Boca Raton, FL, USA.

Kuang, P., Pan, Y., Zhang, Y., et al. (2011). Performance evaluation of a beach nourishment project at West beach in Beidaihe, China. *Journal of Coastal Research* 27(4): 769–783.

Lausman, R., Klein, A.H.F., and Stive, M.J.F. (2010). Uncertainty in the application of the parabolic bay shape equation: Part 1. *Coastal Engineering* 57(2): 132–141.

Morton, R.A., Gibeaut, J.C., and Paine, J.G. (1995). Meso-scale transfer of sand during and after storms: implications for prediction of shoreline movement. *Marine Geology* 126(1–4): 161–179.

National Research Council (1995). Beach Nourishment and Protection. National Academy Press, Washington, DC, USA.

Norcross, Z.M., Fletcher, C.H., and Merrifield, M. (2002). Annual and interannual changes on a reef-fringed pocket beach:Kailua Bay, Hawaii. *Marine Geology* 190(3–4):553–580.

Raabe, A.L.A., Klein, A.H.F., Gonza´lez, M. and Medina, R. (2010). MEPBAY & SMC: software tools to support different operational level of headland-bay beach in coastal engineering projects. *Coastal Engineering* 57(2):213–226.

Ranasinghe, R., Mcloughlin, R., Short, A. and Symonds, G. (2004). The Southern Oscillation Index, wave climate and beach rotation. *Marine Geology* 204(3–4): 273–287.

Roberts, T.M., Wang, P., and Kraus, N.C. (2010). Limits of wave runup and corresponding beach-profile change from large scale laboratory data. *Journal of Coastal Research* 26(1):184–198.

Saravanan, S., and Chandrasekar, N. (2010). Monthly and seasonal variation in beach profile along the coast of Tiruchendar and Kanyakumari, Tamilnadu, India. *Journal of Iberian Geology* 36(1): 39–54.

Short, A.D. (1985). Rip-current type, spacing and persistence, Narrabeen Beach, Australia. *Marine Geology* 64(1–2): 47–71.

Short, A.D. (1996). The role of wave height, period, slope, tide range and embaymentisation in beach classifications: a review. *Revista Chilena de Historia Natural* 69(4): 589–604.

Short, A.D., and Masselink, G. (eds) (1999). Embayed and structurally controlled beaches. In Handbooks of Beach and Shoreface Morphodynamics. Wiley, Chichester, UK, pp. 230–249.

Short, A.D., and Trembanis, A.C. (2004). Decadal scale patterns in beach oscillation and rotation in Narrabeen beach, Australia – time series, PCA and wavelet analysis. *Journal of Coastal Research* 20(2): 523–532.

Short, A.D., Cowell, P.J., Cadee, M., Hall, W. and Van Dijck, B. (1995). Beach rotation and possible relation to the southern oscillation. *Proceedings of the Ocean Atmosphere Pacific Conference*, National Tidal Facility, Adelaide, Australia, pp. 329–334.

Short, A.D., Trembanis, A.C., and Turner, I.L. (2000). Beach oscillation, rotation, and the southern oscillation, Narrabeen beach, Australia. In Proceeding of the 27th International Conference on Coastal Engineering, Sydney, Australia (ASCE). *Coastal Engineering*, Sydney, Australia, vol. 1, pp. 2439–2452.

Silveira, L.F., Klein, A.H.F., and Tessler, M.G. (2010). Headland bay beach planform stability of Santa Catarina State and of the Northern Coast of Sao Paulo State. *Brazilian Journal of Oceanography* 58(2): 101–122.

Stanley Consultants (1985). National Coastal Erosion Study – Final Report. Stanley Consultants, Moffat & Nichol and Jurutera Konsultant (SEA), Kuala Lumpur,Malaysia.

Tan, K.S., Nor Hisham, M.G. and Ong ,H.L .(2005). Coastal Protection against Wave Energy. *Board of Engineers*, Malaysia, pp. 1–19.

van Rijn, L.C. (1998). Principle of Coastal Morphology. Aqua Publications, Amsterdam, the Netherlands. 730p.

Yates, M.L., Guza, R.T., O'Reilly, W.C., and Seymour, R.J. (2009). Overview of seasonal sand level changes on Southern California beaches. *Shore & Beach* 77(1): 39–46.

CHAPTER 6

Impact of a permanent sand bypassing system on the natural sand distribution patterns, southern Gold Coast Australia

The impact of a permanent sand bypassing system on the natural sand distribution patterns around the Southern Gold Coast beaches was investigated. Model investigations were started with the calibration and validation of wave parameters. Wave-induced current patterns were observed during high and low energy wave conditions. Since the model did not include dredging and dumping functionality the discharge of sediments was mimicked using an artificial open channel with the appropriate volume of sediment that could be flushed out by the discharge. Model simulations run by monthly sand bypassing (sand bank discharge) operations were carried out. Results of wave calibration and validation models were in agreement with the results obtained from the field measurement. Comparison of the model results between the case with a sand bypassing operation and the case without the sand bypassing operation showed the former indicates the sand bypassing system has contributed to the additional supplies of sand to the southern Gold Coast beaches. In a case of more sand supply from the bypassing system, the model result showed that a sandspit started to develop at the Rainbow Bay beach. The strong longshore currents push the elongated sandspit to bypass the Greenmount headland and eventually attaching itself to Coolangatta beach. Additional sand supplied by the sand bank discharge operation and obliquely high waves are two main important factors that contribute to the succession of sand bypassing processes around the southern Gold Coast beaches.

Major parts of this chapter were prepared to be published in:
(i) Ab Razak, M.S., Dastgheib, A., and Roelvink, D. (2015). A process-based model of a permanent sand bypassing system. Journal of Coastal Research *(under review).*

6.1 Introduction

Natural sand bypassing is a process where the longshore sand transport along an open coast travels across structures (headland, groyne, and inlet) in the direction of the net sediment transport. Natural sand bypassing across inlets is particularly complex, making it one of the most difficult systems in the coastal environment to quantify. The dominant variables (i.e tidal prism, inlet geometry, wave and tidal energy, sediment supply, spatial distribution of back barrier channels, regional stratigraphy, slope of nearshore, and engineering modifications) controlling the processes and rates of inlet sand bypassing have been documented from numerous case studies (e.g. Fitz Gerald et al., 2000; FitzGerald and Pendleton, 2002; Cheung et al., 2007; Keshtpoor et al., 2013).

For inlets with a tidal prism that is small compared to the alongshore transport rate, a bar will form across the tidal entrance to deliver sand to the down-drift coast. Such bar can be hazardous to navigation. To maintain a navigational entrance and neighbouring beach amenity, engineering modifications at inlets are usually employed and typically involve a combination of training jetties, breakwater and maintenance of a dredge channel. While the result may be an improved entrance channel for a short term period, the training jetties trap the littoral drift in such a way that sand on the up-drift coast accumulates against the training jetty, whilst the down-drift coast erodes due to a lack of sand supply. In the long term, this process may continue until the sand once again naturally bypasses around the entrance, creating another entrance bar (Boswood and Murray, 2001).

An artificial sand bypassing system is a mechanically man-induced transfer of sand from the jetty fillets, shoal, or navigation channel to the down-drift beaches to mitigate the problem associated with the inlet. The sand bypassing system equipped with jet pumps is designed to transfer slurry (a mixture of seawater and sand) from an up-drift beach to a down-drift beach, commonly operated at a jetty inlet. According to Boswood and Murray (2001), the sand bypassing systems can be categorised into three types: (i) water based mobile systems including maintenance dredging either of the channel or sand trap; (ii) land based mobile systems and (iii) permanent (fixed) systems such as trestle or breakwater-mounted.

The most recent sand bypassing system was implemented in 2001 under a project named Tweed River Entrance Sand Bypassing Project (TRESBP), in Australia. This was the second sand bypassing project in Australia after the Nerang River entrance sand bypassing project that has been operated since 1986. Both projects employed a permanent system in which the bypassing plant has a set location. This sand bypassing system requires a high degree of predictability of littoral transport, sediment movement paths and sediment deposition pattern. The challenge poses in the present study is how to model the sand bypassing processes.

In this study, focus is made on the natural sand distribution pattern around the southern Gold Coast beaches due to the impact of permanent Tweed River sand bypassing system. A process based model approach of sand bypassing system is introduced and presented in this study. This study extends the use of process based numerical models to bridge the existing gap of understanding of morphological behaviour of a complex coastal environment. Since the implementation of the permanent TRESBP project, numerous studies related to the operation of the sand bypassing system have been conducted (e.g Castelle et al., 2009; Strauss et al., 2013; Dyson et al., 2001; Castelle et al., 2006; Tomlinson et al., 2007; Acworth and Lawson, 2012). Nevertheless, the use of process based models is limited to long term morphological evolution of Coolangatta bay (Castelle et al., 2009). Understanding of the natural sand distribution patterns due to the impact of a sand bypassing system is a relevant but poorly understood topic. In addition, a process based model that can represent the sand bypassing system has never been developed, so far. Specific objectives are listed out as follows:

(i) to provide an overview of the southern Gold Coast beaches conditions and mitigation measures before and after the construction of the Tweed River permanent sand bypassing system;

(ii) to build a wave-refraction model in order to confirm the validity of offshore wave data collected at the Gold coast wave bouy which is later used as an offshore boundary to morphological models;

(iii) to perform wave calibration and validation tests and comparing measured and modelled data;

(iv) to present a methodological approach of sand bypassing system that is to be applied in the model;

(v) to determine total flow discharges at the inlet discharge boundary that are sufficient enough to flush out the sand bank created in the channel; and

(vi) to investigate the effect of sand bypassing (sand bank discharge) operations on the natural sand distribution patterns around the southern Gold Coast beaches.

The chapter starts with the general description of the natural sand bypassing process, bar formation at the inlet entrance, the applicability of engineering modifications and an artificial sand bypassing system. The chapter continues with the description of study area encompassing the brief history of Coolangatta bay and the operational system of the Tweed River entrance sand bypassing project. The modelling activities begin with the setup of model bathymetry, wave data preparation and analysis of field survey data. The conceptual idea of an artificial sand bypassing system to be implemented in the numerical model was presented and translated into a schematised diagram. The model results were the outputs of the calibration and validation of the wave-refraction model, optimum combination of total flow discharges and sand distribution patterns due to the impact of sand bypassing operations.

6.2 Study area background

6.2.1 Location and setting

The study area is located along the 70 km long Gold Coast, in Queensland Australia. The area of investigation covers approximately 6 km of coastline, comprising three distinct embayments within Coolanggatta Bay: Rainbow bay beach, Coolangatta beach, and Kirra beach (see **Figure 6-3**). The study areas are characterised by the Tweed River entrance and a major headland called the Point Danger which is located at the most southern end of the Gold Coast. Another prominent feature is a small headland i.e Greenmount Hill (see **Figure 6-1 (a)**) that separates the Rainbow bay beach and the Coolangatta beach. Towards the west, structural groynes named Kirra Point (see **Figure 6-1 (b)**) and Mile Street are placed on the beach, functionally designed to block longshore drift in order to provide wider beach area at Coolangatta. Kirra beach starts from the Kirra Point groyne towards the west (see **Figure 6-1 (c)**).

Figure 6-1. Coolangatta bay beach which is bounded by the (a) Greenmount Hill headland and (b) Kirra Point groyne and (c) Kirra beach, a view from the Kirra Point.

Prior to the implementation of the Tweed River sand bypassing system, Kirra beach had suffered from high energy waves and a reduced sand supply. As a consequence, the beach at Kirra is severely eroded. **Figure 6-2 (a)** shows the Kirra beach condition a year before the installation of the permanent sand bypassing system. The rocky area behind the Kirra Groyne Point is exposed and the short Mile street groyne is still visible. After the installation of the bypassing system, Kirra beach grows healthily. The beach is wider than before and the Mile Street groyne is hardly seen as it is covered by sand. Sand bypasses the Kirra Groyne Point and fills the exposed rocky area as seen in **Figure 6-2(b).**

Figure 6-2. Kirra beach condition in (a) May 2000 and (b) in April 2012.

The study area is exposed to high energy waves. The Southeast Queensland region is characterised by wave dominated coasts and its wave climate is subjected to important variances. South to southeast waves are generated by intense low pressure systems off the New South Wales coast in winter and spring with average deep-water significant wave heights ranging from 0.8 to 1.4 m and mean wave periods of 7 to 9 s. This contributes to the main component of the northerly longshore drift. Nevertheless, south-easterly swells do not have a strong impact on the Gold Coast beaches, particularly the southern part of the Gold Coast beaches which are sheltered from this direction and, in addition, draw advantage from an artificial sand bypassing system (Dyson et al., 2001). Tropical cyclones that commonly occur from December to April can generate northeast to east waves, sometimes with a destructive power (Hobbs and Lawson, 1982), with a significant wave height up to 8m. This is supported by the recent study of Castelle et al., (2007) where high energy easterly swells have a considerable destructive power on both northern and southern Gold Coast beaches. Tides are semi-diurnal, with a tidal range varying from 0.2 to 2.0 m, with a mean of 1m. The sediment consists of fine sand, with $D_{50} = 200$ μm. The estimated net rate of littoral sand transport within the general Gold Coast region is of the order of 500,000 m³/year toward the north (Turner et al., 2006). The Tweed River itself is considered to be a net sink for sediment, as the river only discharges minor quantities of fluvial sand to the littoral system.

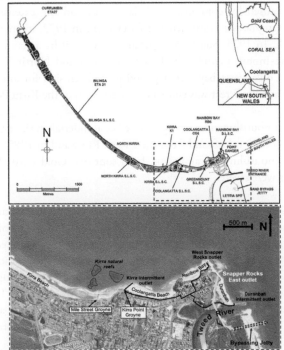

Figure 6-3. Location and general setting of Coalanggata Bay comprising of Rainbow bay beach, Coolangatta beach, and Kirra beach. The Tweed river entrance and the layout of the permanent sand bypassing system are shown. Upper figure is courtesy from Strauss et al., (2013) and lower figure is courtesy from Castelle et al., (2009).

6.2.2 History of Coolanggatta Bay prior to TRESB project

In 1960s, the Tweed River entrance training jetty was extended approximately 380 m to improve the navigability of the river entrance. Navigation conditions improved as a result of the engineering works, but this improvement did not last. The jetty extension improved navigation condition for almost 20 years. Sand accreted to the south of the entrance and as sand began to pass the entrance again, a new bay formed and navigation condition worsened (Tomlinson et al., 2007)

The extension of the Tweed River entrance training jetty reduced the net northerly transport of sand moving to Queensland and resulted in severe erosion, recession and vulnerability of the southern Gold Coast beaches to large swell events. The loss of longshore sand supply from the south resulted in progressive recession of the Coolangatta bay beaches. Panel A and panel B in Figure 6-4 show conditions of Coolangatta bay before and after the extension of Tweed River training jetty.

In addition, severe storms in the late 1960s and early 1970s resulted in major erosion along the entire south east Australian coastline. A big storm in 1967 for instance, further reduced the volumes of sand on the beach and potentially exposed coastal property to greater risk. As a result of ongoing erosion caused by combination of large seas and the extension of Tweed River training jetty, a large groyne was constructed at Kirra Point in 1972, in an attempt to hold sand on Coolangatta beach. In 1974, South east Queensland was hit by a series of cyclones and east coast lows over a five-month period. The beaches at Kirra and North Kirra were completely eroded and put the Coolangatta bay into a period of high erosion for almost a generation. In 1975, a small groyne at Mile Street was constructed westwards of the Kirra Point groyne.

A number of beach nourishment campaigns have been undertaken since then. In 1975, a trial project consisting of 76,000 m³ of sand from the Tweed River and in 1985 followed by 315,000 m³. Despite previous nourishment campaigns, both southern Gold Coast beaches erosion and navigation conditions were severe.

Prior to permanent sand bypassing system, a series of nourishment works have been undertaken under the Tweed River Entrance Sand Bypassing (TRESB) project. The first major nourishment that took place between November 1989 and May 1990 is called the southern Gold Coast Beach Nourishment Project. A total of 3,200,000 m³ of sand was placed on the nearshore and upper beach extending from Kirra East to several kilometers north using sand sourced from inactivate offshore deposits in water depths of 20 to 28 m. By the early 1990s, North Letitia Beach, located immediately to the south of the southern training wall, had accreted significantly that a sub-tidal delta had once again formed at the Tweed River entrance, creating a navigational hazard for vessels.

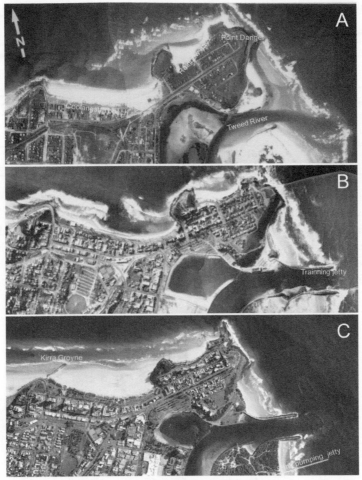

Figure 6-4 . Evolution of Coolangatta bay (A) Tweed River entrance in 1935, before the Tweed River training jetty were extended (B) Tweed River entrance in 1967 shortly after the Tweed River jetty were extended (C) Tweed River entrance in 2004 three years after the bypass system was commissioned (Acworth and Lawson, 2012)

6.2.3 TRESB project and sand bypassing system overview

The TRESB project, which was started in 1995 is a joint coastal engineering project of the New South Wales and Queensland State Governments. This project is formulated to overcome both the significant erosion of the southern Gold coast beaches and the navigation issues due to the Tweed River entrance infilling. The project comprises of two nourishment stages i.e Stage 1 and Stage 2. **Table 6-1** summarises the dredge quantities of Tweed River entrance and associated nourishment (Castelle et al., 2009) of the southern Gold Coast beaches over a period of 19 years.

Table 6-1: Dredging quantities and associated nourishment over the period 1995 to 2014.

Periods	Stages	Dredging quantities/ Nourishment volume (m³)	Activities
1995-1996	Stage 1A	2,300,000	Dredging of tweed River entrance
1997-1998	Stage 1B	800,000	and nourishment of southern Gold Coast
2000-2002	Stage 2A	1,100,000	beaches
2001	Stage 2A		Start of the permanent sand bypassing system
2003-2006	Stage 2B	500,700	Dredging of tweed River entrance and nourishment of southern Gold Coast beaches
2007-2014	Stage 2B	199,179	Dredging of tweed River entrance and nourishment of Duranbah beach

Stage 1 involved removing sand bars from the Tweed River entrance. The sand was placed on the upper beaches from Rainbow bay in the east to North Kirra in the west. Additional sand quantities were placed in the nearshore. Stage 2 resulted from refinements to the Stage 1 placement areas (Boswood et al., 2001 and Colleter et al., 2001). Most of the sand was placed in an area to the east of Snapper Rock. **Figure 6-5** shows the dredging area and associated nourishment areas.

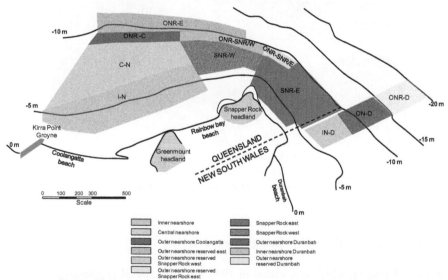

Figure 6-5. Tweed River entrance dredging area and associated nourishment areas.

The permanent sand bypassing system which was installed in 2001 is an innovative aspect of the TRESB project. The idea of the permanent sand bypassing system is to collect sand from the southern side of the Tweed River entrance and transport it to the southern Gold Coast beaches for an indefinite period of time. Five locations (Duranbah outlet, Snapper Rock East outlet,

Snapper Rock West outlet, Greenmount outlet, and Kirra outlet) within Coolangatta bay were designated for the pumping outlets as can be seen in **Figure 6-6.**

Figure 6-6. Tweed River entrance training jetty and locations of outlet sand pumping. Red arrows indicate sand pumping directions and yellow arrows shows net littoral longshore sand transport.

The permanent sand bypassing system has 11 jet pumps, which operate at depths of 6.5 to 15 m below the mean sea level. The seabed level at the end of the pier at commissioning was about 6 m below mean sea level. Each jet pump uses a 50mm diameter, 35 to 50 m/s water jet to entrain the surrounding sand and water mixture. This mixture is thickened in a slurry pit to a specific gravity of about 1.4 and then pumped via a 400 mm diameter polyurethane lined steel pipe to the outlets. The system's mechanical capacity is about 1500 m3 of slurry per hour, which is equivalent to about 500 m3 of sand per hour. The sand bypassing system was installed in January 2011 and the first slurry was pumped in February 2001 followed by significant pumping in March 2001. The pumping slurry was pumped to the Snapper Rocks East outlet (main outlet).

Over the period 2001 to 2013, the majority of sand was pumped to the primary Snapper Rocks East outlet, at the extremity of Coolangatta bay. Sand was frequently pumped to the Duranbah beach but in fewer quantities than the pumping capacities at the primary Snapper Rocks East outlet. **Table 6-2** shows the yearly pumping quantities at five different outlet locations over the period 2001- 2013. The yearly total sand pumping quantities are comparable to the natural longshore drift (~500,000 m³/year) in order to allow the permanent sand bypassing system to work in concert with natural processes.

Table 6-2: Sand pumping quantities (m³) over the period 2001 to 2013.

Year	SR-E	SR-W	Kirra	Duranbah	Greenmount	Total
2001	409,746	1,586	97,279	67,258	0	575,869
2002	564,577	0	84,915	71,872	0	721,364
2003	602,953	0	81,918	70,224	31,931	78,7026
2004	459,554	0	0	36,813	0	496,367
2005	683,244	0	0	41,687	0	724,931
2006	485,185	0	0	67,099	0	552,284
2007	514,968	0	0	47,279	0	562,247
2008	520,312	0	0	65,497	0	585,809
2009	354,743	10,678	0	24,211	0	409,232
2010	373,828	0	0	21,781	0	359,609
2011	457,765	0	0	60,404	0	518,169
2012	354,910	0	0	57,482	0	436,092
2013	309,279	0	0	10,604	0	319,883

SR-E　　　　: Snapper Rocks-East primary outlet.
SR-W　　　　: Snapper Rocks-West intermittent outlet.
Kirra　　　　: Kirra intermittent oultet.
Duranbah　　: Duranbah intermittent outlet.
Greenmount : Greenmount intermittent outlet.

6.3 Wave data

6.3.1 Gold Coast offshore wave data

The offshore wave directional data were collected at a Gold Coast bouy located at a 18 m water depth. This half-hourly offshore wave data was sourced by the Gold Coast City Council (GCCC) covering the whole year of 2011 as can be seen in **Figure 6-7**. Waves predominantly come from the southeast to east directions with the significant wave height ranging from 0.5 to 6.0 m in both years. For the 2011 annual wave data, 45 percent of waves are in a range of 0.5 to 1.0 m followed by 33 percent that have wave heights of 1.0 to 1.5 m. Waves that exceed 2.0 m height account for 7 % of the total waves. Wave periods are relatively long due to the nature of swell waves.

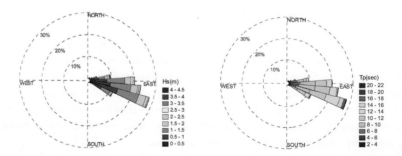

Figure 6-7. Annual offshore wave roses (2011) at the Gold Coast obtained from the Gold Coast wave data directional bouy.

6.3.2 Nearshore wave measurement

Data from a field measurement campaign along five Gold Coast beaches (i.e. Kirra, tugun, Palm Beach, Burleigh and Narrowneck) was kindly provided by the Gold Coast City Council (GCCC).**Figure 6-8** shows the measurement locations where the data were collected. Based on the technical report of the Gold Coast Shoreline Management Plan (Field measurement and data collection) which was written by Stuart and Lewis (2011), four main data were required by the GCCC to be collected including (i) nearshore wave-current measurement within the surf zone along the Gold Coast coastline (ii) tidal current measurements at the mouth of Tallebudgera and Currumbin Creeks (iii) water level measurements within Tallebudgera and Currumbin Creek estuaries and (iv) Langarian current maps using GPS enabled drogues.

Due to limitations on the use of this data, only nearshore wave measurement is considered to be reported and point of measurement locations are restricted to Kirra beach and Tugun beach only. This is because the area of interest is restricted to the southern Gold Coast beaches only. **Table 6-3** shows the deployment details of nearshore ACDPs.

Table 6-3: Deployment details of ACDPs

Location	Easting	Northing	Depth (m)	Measurement periods
Kirra	553249	6884852	6.8	12.4.11 - 3.5.11
Tugun	549102	6887198	7.3	9.5.11 - 30.5.11

Figure 6-8. Field measurement locations

Acoustic Doppler Current Profilers (ACDPs) were deployed at Kirra and Tugun locations collecting 6 weeks' worth of wave and current data in the nearshore zone in a water depth of approximately 7.0 m. The ACDPs were first installed on 12.4.2011 and set to record a 10 minutes burst of data every hour. The bin size was set at 0.5 m with the first bin range of 1.62 m. On 4.5.2011, the instruments were removed from the water, cleaned, data were downloaded and new batteries were installed. Due to bad weather conditions, the instruments were re-deployed on 9.5.2011 at the same positions and further data collection period the instruments were retrieved on 30.5.2011.

Figure 6-9 shows the nearshore wave parameters at two locations which were measured by the ACDP. At both locations, the measured significant wave heights do not differ too much. The wave heights range from 0.5 to 2.5 m at both sites. The dominant wave directions at Tugun however are considerably higher than at Kirra. Waves mostly propagate easterly and east-southerly at Kirra and Tugun, respectively. At both locations, the peak wave periods consist of both short and long periods of waves. Similar to wave height, the peak wave periods at both locations do not differ too much.

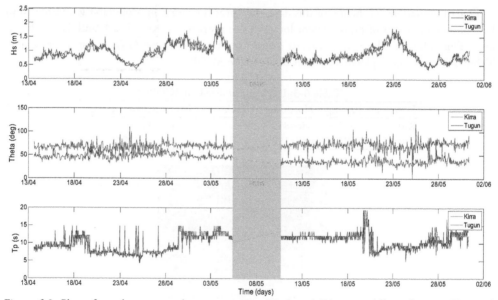

Figure 6-9. Plots of nearshore measured wave parameters (H_s, θ, and T_p) at two different locations Kirra and Tugun. Shaded area denotes periods of no measurements.

6.4 Model setup

6.4.1 Grids and initial bathymetry

A curvilinear grid meshes were used with varying grid sizes (dx = 5 to 30 m, dy = 10 to 20 m), coarser grid cells near the offshore boundary and finer grid cells at a nearshore zone. The curvilinear grids were used in this modelling work because the study areas are generally characterised by curve beaches, surrounded by oblique structures and natural headlands. A refined curvilinear grid system is suitable to represent this type of a unique coast and complex environments. The grids were extended further to the north part of Tugun and south part of Letitia Spit in order to allow sufficient coverage of the propagated waves into the area of interest.

A bathymetry survey data of March 2011 was used as an initial bathymetry for the calibration and validation of wave models as well as for the morphological model. Topographic data especially the area of headlands was artificially created. The Tweed River itself was treated as land in this study. Along the offshore boundary, depths are uniformly set at 25 m. **Figure 6-10** shows the grid and bathymetry of the model.

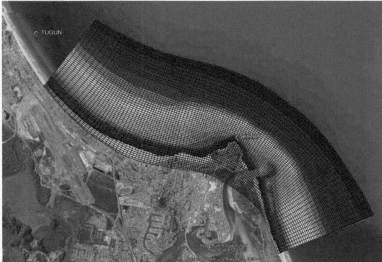

Figure 6-10. Grid and bathymetry plots at Coolangatta bay. The bathymetry survey was undertaken in 2011 and 2012. The coarse grid is shown for visibility. Finer grids (refined by a factor of 9) are used for the model computation. Shaded map ranges from 0 m (blue) to -25 m (red) depths.

6.4.2 Offshore wave model boundary

A set of calibration-validation models was built to observe the wave refraction pattern around the Coolanggatta bay. These numerical modelling exercises are important to ensure that waves are properly propagated into the model domain as well as to confirm the validity of the Gold Coast offshore wave data. For the calibration and validation purposes, offshore waves data for a selected period were used. **Figure 6-11** shows the plot of offshore wave parameters during the calibration and validation periods. The offshore wave includes both low and high energy waves which cover a period of 30 days.

During the calibration periods (from 20/4/11 to 28/4/11), the offshore significant wave heights range from 0.5 to 2.0 m, associated with the wave periods from 5 to 10 s. After day 28, the significant wave height rises up to 2.0 m and in some days in May waves can go beyond 2.0 m height. It is expected that within the periods where the measurement are not taken, wave can be persistently high due to bad weather conditions. The same phenomenon was observed for the peak wave period as can be seen in **Figure 6-11 (middle panel)** where the wave period is greater than 10 s beyond day 28. The offshore wave direction as plotted in **Figure 6-11 (lower panel)** is not affected by the variation of wave height and wave period. The offshore wave propagates from east to east-south easterly directions.

During the validation periods (10/5/11 to 26/5/11), the offshore wave heights are moderate and relatively low compared to the wave heights during the calibration periods. Waves are dominant from the east.

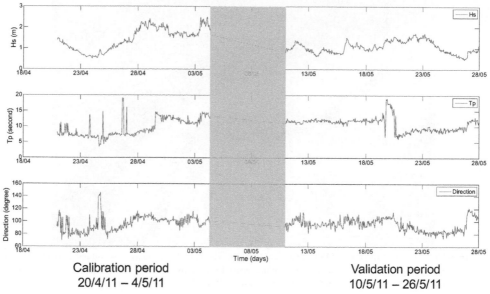

Calibration period	Validation period
20/4/11 – 4/5/11	10/5/11 – 26/5/11

Figure 6-11. Plot of offshore significant wave height, wave period and wave direction as an offshore wave boundary condition. Shaded area denotes periods of unavailable nearshore measurement.

6.4.3 Wave calibration model setup

Two main wave parameters in XBeach were tuned to observe the sensitiveness of these coefficients to measured wave parameters. XBeach uses a modified form of the parametric (empirical) dissipation formulation of Roelvink (1993) by default. **Table 6-4** shows the calibration model parameters of XBeach where the wave breaker indexes ranging from 0.35 to 0.65 (0.55= default in XBeach) while the short wave friction coefficients in the order of 0.0 to 0.15 (0.0= default in XBeach).

For the wave spectrum boundary condition, a parametric Jonswap were applied. New waves were called every 30 minutes and the wave time step of 1.0 s was used to generate a random spectrum across the sea. Gammajsp (γ_{jsp}) was selected as 1.0 to represent the nature of swell wave which is common for the Gold Coast. All models were run for 14 days beginning 20/4/11 and ending 4/5/11. Model outputs were directly compared with data measurement.

Table 6-4. Wave calibration model parameters

No.	Model parameters	Values
1.	Upper directional limit (nautical in deg.)	160
2.	Lower directional limit (nautical in deg.)	20
3.	Directional resolution (deg)	20
4.	Wave breaking formula	roelvink3
5.	Wave breaker index, γ	0.35 : 0.45 : **0.55** : 0.65
6.	Short wave friction, f_w	**0.0** : 0.15 : 0.30 : 0.45
7.	Simulation time	14 days

Bolded items are default values in Xbeach

6.4.4 Wave validation model setup

Model validation was done by applying the optimum combination of short wave friction and wave breaking coefficients that were achieved from the calibration tests. A validation model was run from 10/5/11 to 26/5/11 covering a period of 16 days. Validated model outputs were compared with measurement data.

6.4.5 Statistical model evaluation (Error index)

Two statistical models were used to evaluate the performance of XBeach model in simulating wave refraction pattern within the Coolangatta bay. Both statistical methods, which were adopted from Moriasi et al., (2007) are known as percent bias (PBIAS) and RMSE-observations standard deviation ratio (RSR).

PBIAS measures the average tendency of the simulated data to be larger or smaller than their observed counterparts. The optimal value of PBIAS is 0.0, with low magnitude values indicating accurate model simulation. Positive values indicate model underestimation bias and negative values indicate model overestimation bias. PBIAS is calculated by **Equation 6.1** where PBIAS is deviation of data being evaluated, expressed in percentage i.e.

$$PBIAS\ (\%) = \left[\frac{\sum_{i=1}^{n} (Y_i^{obs} - Y_i^{sim}) * (100)}{\sum_{i=1}^{n} (Y_i^{obs})} \right]$$

(6.1)

RSR standardise root mean square error (RMSE) using the observation standard deviation and it combines both an error index and the additional information recommended by Legates and Mc Cabe (1999). RSR is calculated as the ratio of the RMSE and standard deviation of measured data as derived in **Equation 6.2**.

$$RSR = \frac{RMSE}{STDEV_{obs}} = \frac{\left[\sqrt{\sum_{i=1}^{n}(Y_i^{obs} - Y_i^{sim})^2}\right]}{\left[\sqrt{\sum_{i=1}^{n}(Y_i^{obs} - Y_i^{mean})^2}\right]} \tag{6.2}$$

RSR varies from the optimal value of 0.0, which indicates zero RMSE or residual variation and therefore perfect model simulation, to a large positive value. The lower RSR, the lower the RMSE and the better the model performance. **Table 6-5** shows the statistical model performance rating based on Moriasi et al., (2007).

Table 6-5: General model performance rating

RSR (-)	Model performances	PBIAS (%)
<0.5	Very Good	<10
0.5-0.6	Good	10-15
0.6-0.7	Satisfactory	15-25
>0.7	Unsatisfactory	>25

6.4.6 Morphological bed patterns without bypassing

Once the model calibration and validation are done, the next step is to investigate the sand distribution patterns and morphological bed changes. The grid and bathymetry setups that were used for the calibration and validation models were re-used for the morphological model. A sediment transport formulation of Van Thiel-Van Rijn was used to compute the magnitude of bed and suspended sediment transport. Sediment grain size of 200 μm was distributed across the whole domain. Headland and groyne were represented as non-erodible (hard) structures. A 10 m erodible layer of sand is allowed to excite for any changes on the sea bed. Sediment transport behaviour and near shore morphological bed patterns were observed during the simulation periods *i.e* from 20/4/2011 to 25/5/2011.

6.4.7 Schematisation of sand bypassing system

For the scenario with the inclusion of permanent sand bypassing, an artificial sand bypassing system was created. In Xbeach, this can be done by utilising a river discharge option. In our model, the sand bypassing system consists of a long fixed river channel and a sand bank. The river channel is artificially created running from a land model boundary to the nearshore area where the sand is supposedly discharged to the sea (see **Figure 6-12(a)**). Inside the river channel, an artificial sand bank is created and it is treated as an erodible structure as can be seen in **Figure 6-12(b)**. Whenever there is a sufficient flow discharge supply from the inlet boundary, the sand bank will be eroded and carried away out of the channel by the flow current. The volume of sand bank should be equal to the volume of sand that is pumped out through the outlet (see **Table 6-**

6). A series of tests is preliminary carried out to determine the optimal total flow discharges that are able to flush the sand bank out to the sea within a period of one month.

The presence of the Tweed River is neglected in this numerical modelling study. The river itself is treated as a land and therefore the river discharge is simply ignored. The sand shoal that is naturally developed in front of the Tweed River jetty is preserved and is treated as an erodible structure. This means the sand bypassing at the Snapper Rock may also have an influence from the sand supplied by the sand shoals at the Tweed River jetty but the contribution is expected to be minimal.

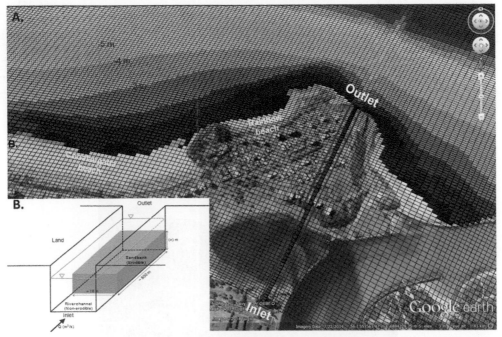

Figure 6-12. (a) Plan view of an artificially created river channel (b) Three dimensional view of sand bank inside the river channel.

Table 6-6 shows the monthly sand pumping quantities starting from January to December 2011. Pumped sand was mainly delivered to the main outlet of Snapper Rock-East (see **Figure 6-13**) and sand was occasionally pumped to Duranbah outlet. A total of 518,619 m³ volume of sand was successfully pumped out to these two outlets. The total volume most likely represents the amount of natural longshore transport i.e 500,000 m³/year (Turner et al., 2006). In this study, the presence of the other four outlets (Snapper Rock-west, Kirra, Duranbah, and Greenmount) is ignored and focus is made on the main outlet (Snapper Rock east) only. On the other hand, the morphodynamic simulations were performed beginning of April 2011 and ending December 2011. This is because the initial bathymetry was surveyed somewhere in March 2011. Therefore, the total volume of the pumped sand is less i.e 348,109 m³/year.

Table 6-6: Monthly sand pumping quantities

Pumping quantities (m³) from January 2011 to December 2011					
Month	SRE	SRW	Kirra	Duranbah	Greenmount	Total
January	35,722	-	-	-	-	35,722
February	23,080	-	-	6,460	-	29,540
March	50,854	-	-	-	-	50,854
April	45,325	-	-	-	-	45,325
May	27,031	-	-	24,674	-	51,705
June	63,363	-	-	-	-	63,363
July	53,621	-	-	-	-	53,621
August	56,251	-	-	-	-	56,251
September	37,325	-	-	-	-	37,325
October	20,206	-	-	24,729	-	44,935
November	5,174	-	-	4,541	-	9,715
December	39,813	-	-	-	-	39,813
					Total	518,169

*SRE: Snapper rock east outlet (main outlet); SRW: Snapper rock west outlet

Figure 6-13. An aboveground pipe discharges sand at the Snapper Rock east outlet.

6.4.8 High performance cloud (HPC) computer

All model simulations performed in this paper were successfully done by using high performing cloud (HPC) facilities. The HPC facilities were provided by surfsara, a company that managed the use of high performance computers with the aim to support research developments in the Netherlands. The HPC enables a Message Passing Interface (MPI) approach that allows several core computers to work in a parallel mode. The use of MPI was tested in both Windows and Linux computers. Based on the tested runs, models that run in Linux based operating system computer are significantly faster than the models that run in Windows based operating system computer.

6.5 Results and discussions

6.5.1 Wave calibration

Effect of short wave friction (f_w) coefficients

Models were run with four different coefficients of short wave frictions while the breaking wave parameter (γ) was held constant at 0.55 for all four models. **Figure 6-14** shows comparison of measured and modelled wave parameters at two different sites i.e Kirra and Tugun due to the effect of short wave friction (f_w) parameters. The f_w parameter influences the measured significant wave heights at both site locations. The higher the f_w values, the greater the difference between the measured and modelled significant wave heights. The absence of f_w ($f_w = 0.0$) resulted in a close agreement between the measured and modelled wave heights. In this study, the short wave friction is not necessarily needed as the bottom bed bathymetry is entirely smooth. The larger f_w value i.e $f_w > 0.0$ are applicable for coastal environments where the coral reef presents. For instance, van Dongeren et al., (2012) suggested a f_w value of 0.6 for modelling wave and currents across a fringing coral reef.

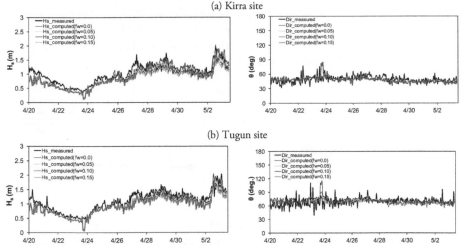

Figure 6-14. Comparison between measured and modelled wave parameters (H_s and θ) for varying wave friction parameters

Table 6-7 shows the skill performance of the models with varying f_w. The PBIAS shows the increment of the bias percentages as the f_w increases at both Kirra and Tugun sites. This indicates higher f_w leads to large deviation of measured and modelled significant wave heights. Nevertheless, the RSR skill shows a very good performance of XBeach model in all cases, despite the increment in PBIAS. This infers RSR is defined based on the results' trend-line.

On the other hand, short wave friction coefficient does not influence the measured wave direction. At both sites, results of measured and modelled wave directions are coincided regardless of the f_w coefficients. Both PBIAS and RSR skills show a very good performance of the XBeach model.

Table 6-7: Model skills performance (short wave friction)

F_w	PBIAS (%)				RSR (-)			
	H_s		θ		H_s		θ	
	Kirra	*Tugun*	*Kirra*	*Tugun*	*Kirra*	*Tugun*	*Kirra*	*Tugun*
0.0	4.33	7.71	1.30	-4.08	0.06	0.07	0.01	0.23
0.05	9.86	10.48	2.17	0.75	0.11	0.20	0.11	0.04
0.10	13.95	12.95	2.59	0.88	0.16	0.01	0.10	0.20
0.15	17.43	15.23	2.69	1.22	0.19	0.01	0.10	0.19

Effect of breaking wave (γ) coefficients

A separate set of four model simulations was then conducted, this time using a constant short wave friction (f_w=0.0), but for four different values of the breaking coefficient γ(0.35,0.45,0.55,0.65). **Figure 6-15** shows a comparison of modelled and measured wave parameters at Kirra and Tugun sites due to the effect of varying breaking wave coefficients. In all different wave breaking coefficient cases, the γ does not influence the measured wave parameters neither the significant wave height (H_s) nor the wave direction (θ). This is due to the fact that propagated waves still do not break as a breaking condition (H_{rms}>γh) in a depth-limited breaking model of XBeach does not yet satisfied.

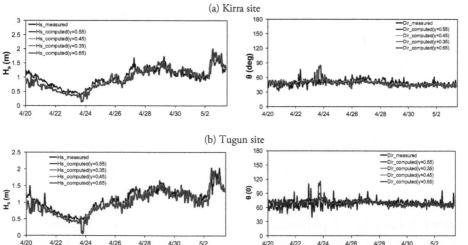

Figure 6-15. Comparison results between measured and modelled wave parameters (H_s and θ) for varying breaking wave coefficients

Table 6-8 shows the skill performance of the models with varying γ. PBIAS skills of H_s for Kirra and Tugun sites fall between good and very good performances. Likewise, PBIAS skills of θ for both sites indicate a very good performance; although model results for γ=0.55 over-predict measurement. RSR skill results for all breaking wave coefficients show a very good performance of the XBeach model.

Table 6-8: Model skills performance (breaking wave parameter)

γ	PBIAS (%)				RSR (-)			
	H_s		θ		H_s		θ	
	Kirra	*Tugun*	*Kirra*	*Tugun*	*Kirra*	*Tugun*	*Kirra*	*Tugun*
0.35	5.35	7.84	1.49	0.86	0.07	0.13	0.11	0.19
0.45	5.20	7.83	1.46	0.13	0.07	0.86	0.11	0.19
0.55	4.33	7.78	1.30	-4.08	0.06	0.13	0.01	0.23
0.65	5.10	7.82	1.41	0.85	0.07	0.13	0.11	0.19

* Kir=Kirra; Tug =Tugun

6.5.2 Wave validation

Based on the model calibration results, a combination of f_w=0.0, and γ=0.55 is found to be the optimum parameters for a validation test. This set of optimum parameters was subsequently used to simulate the entire wave event from May 10 00:00 hours, to May 26 23:00 hours (16 days in total).

Figure 6-16 shows a comparison between the results of modelled and measured wave parameters at Kirra site together with the plots of hourly offshore wave parameters which were imposed at the offshore model boundary. Offshore significant wave height reduces significantly as it approaches to shallower water as can be seen in **Figure 6-16 (upper panel)**. A good agreement is generally observed throughout the simulation. However, the result of modelled H_s slightly underestimates measured H_s during the last four days of simulation.

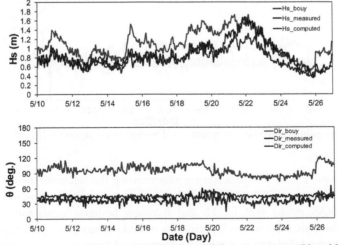

Figure 6-16. Comparison results between measured and modelled wave parameters (H_s and θ) at Kirra site.

During the last four days, the measured H_s are consistent with the offshore H_s which is quite surprising. From this it can be inferred that there is no change in wave energy during these periods despite significant wave direction changes from offshore to nearshore. The modelled wave heights consistently reduce over the course of simulations with an average reduction of 29 % relative to offshore wave heights. The modelled wave heights pattern is therefore considered reasonably predicted. Wave direction changes significantly over the 16 days simulation as can be seen in **Figure 6-16** (bottom panel). Easterly swell direction changes to north-easterly (40⁰ with respect to shore normal) when waves arrive at depth of approximately 7 m. Modelled wave directions are in a good agreement with the measured data, although they remain consistently over-predicted.

Figure 6-17 shows a comparison between the results of measured and modelled wave parameters for Tugun site. The modelled significant wave heights are fairly predicted against measured wave heights with a slight underestimation during the last 10 days of simulation. At Tugun, wave refracted quite obliquely due to the orientation of the coast. The coastline at Tugun has an angle of approximately 40⁰. When offshore easterly waves propagate toward Tugun beach, they attenuate and refract at an angle of 60⁰. Overall, modelled wave directions are generally in a good agreement with the measured wave directions.

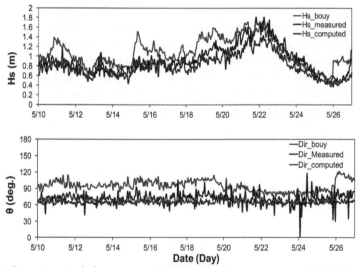

Figure 6-17. Comparison results between measured and modelled wave parameters (H_s and θ) at Tugun site.

Table 6-9 shows model skill performance values during the validation periods at both Kirra and Tugun sites. In general, XBeach model did a good job in predicting wave measurement data. Despite the PBIAS model skill for H_s and θ ranges between satisfactory and very good performances, the RSR skills for both parameters indicate a very good performance inferring a good trend of the modelled data in comparison to measured data.

Table 6-9: Model skills performance (optimum calibrated parameter)

Sites	PBIAS (%)		RSR (-)	
	H_s	θ	H_s	θ
Kirra	9.91	-23.19	0.08	0.16
Tugun	12.60	8.80	0.02	0.13

6.5.3 Wave-induced surf zone current patterns

Coolangatta bay is characterised by an intense longshore current all year long, under sufficient offshore wave conditions. This longshore current is responsible for vigorous sand transport along the southern Gold Coast beaches. Additionally, large amount of sand available updrift of Snapper Rock supplied by the monthly sand bypassing system is the main contributor to significant and rapid morphological evolution of the Coolangatta bay. In this contribution, the modelled wave-induced current patterns under high and low wave energy conditions during the validation periods are investigated.

Figure 6-18 shows the computed wave-induced current patterns along the southern Gold Coast beaches for offshore wave conditions: H_s = 1.73 m, T_p= 9.77s and θ=86^{0}N occurs on May, 22 2011. This offshore wave condition contains the highest significant wave height which was recorded by the Gold Coast offshore wave bouy during the validation period. This simulation result shows a predominant westward longshore current along the bay. This longshore current is also accelerated in front of Greenmount Hill headland and Kirra groyne (invisible due to sand accretion). The wave induced current magnitude reaches 2.0 m/s near the Snapper Rock headland and the Rainbow bay which is a quite significant intensity given H_s 1.4 m at the breaking at Snapper Rock.

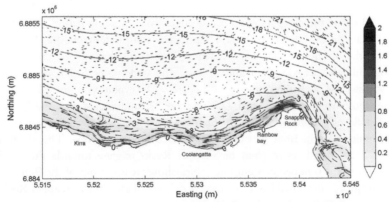

Figure 6-18. Modelled wave-induced current magnitude (m/s) along the southern Gold Coast coasts for offshore wave conditions: East swell (87^0 N) with H_s = 1.7 m and T_p= 9.8 s. Colour bars indicate velocity speed (m/s)

Figure 6-19 shows the modelled wave-induced current pattern along the southern Gold Coast beaches for a low offshore wave condition: H_s = 0.5 m, T_p= 9.3s and θ=78^{0}N occurs on

May, 25 2011 during the validation period. During the low wave energy event, some counter-clockwise circulation cells are observed in front of the Rainbow bay beach and Coolangatta beach. The formation of these small circulations is due to the presence of an irregular bathymetry. Westward longshore currents as observed during the high wave energy exist but are less significant. At Snapper Rock, longshore currents are accelerated by easterly waves with the current magnitude of merely 0.8 m/s.

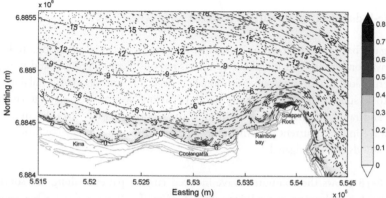

Figure 6-19. Modelled wave-induced current magnitude (m/s) along the southern Gold Coast beaches for offshore wave conditions: Easterly swell (78^0 N) with H_s = 0.5 m and T_p= 9.3 s. Colour bars indicate velocity speed in m/s.

6.5.4 Morphological bed change patterns (without sand pumping operation)

The morphological run was conducted starting from 20/4/2011 and ending 27/5/2011, covering both calibration and validation periods. During these periods, sand pumping operation was presumably inactivated.

Figure 6-20 shows the morphological bed changes over the study period. In general, model results show that the easterly waves have a significant contribution on the sand distribution patterns along the southern Gold Coast beaches. The morphological bed changes apparently occur in water depths between 0 m and -4 m. The underwater bed contour lines change immediately as soon as the bed morphology is updated. The evolution of these bed contour lines indicates the sand movement on the bed. During these 37 simulation days, easterly waves largely transport a significant volume of sand at the Snapper Rock headland towards the west side of the coast. As a result, areas in front of the Snapper Rock headland erode. The -1 m and -2 m underwater bed contour lines in front of Snapper Rocks migrate towards the headland and eventually merge to the 0 m bed contour line. Nearshore areas in front of the Rainbow bay and Coolangatta are greatly benefited from the massive delivery of sand at the Snapper Rock.

6. Impact of a permanent sand bypassing on the natural sand and distribution patterns, southern Gold Coast beaches, Australia

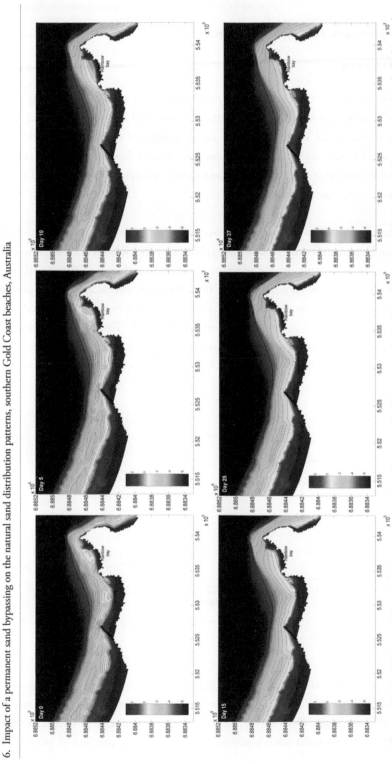

Figure 6-20. Modelled morphological bed changes during the calibration and validation periods. In all panels iso-contours (0.5 m intervals) are contoured in the background.

6.5.5   Impact of a permanent sand bypassing system

Total flow discharge rate

The most important parameter in determining the succession of the permanent sand bypassing through the numerical modelling of this study is the total flow discharge rate. The total flow discharge in the model ideally represents the speeding rate of the mechanical sand bypassing pump. A set of models with different combinations of total flow discharges was run to determine the total flow discharge rate that is sufficient enough to fully remove the sand bank out of the channel within the one month period.

In a case when the same total flow discharge is applied to both models that were run with and without wave, the results were different. This can be explained by the additional contribution of the wave induced longshore currents. In a case when the wave is included in the model, the interaction between the longshore current and the flow discharge currents occurred at the channel's entrance slowdown the removal process of the sand bank. As a result, residual sand still remains trapped inside the channel as can be seen in **Figure 6-21**. The low magnitude of discharged flow currents was not strong enough to push the sand bank away from the channel. Therefore, model with the inclusion of waves requires higher total flow discharge that the model without waves.

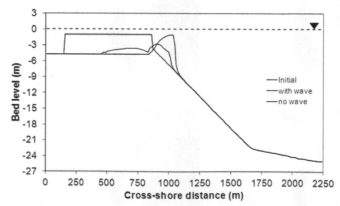

Figure 21. Modelled cross-shore sand bank profiles for two different cases i.e. simulation with wave and simulation without wave. The total flow discharge rate is held constant for both simulations.

Table 6-10 shows the results of the optimal flow discharges both for models run with wave and models run without wave for different total flow discharges. The flow discharge rate is divided into three different stages. From day 0 to day 10, the flow discharge is considerably small, constant at 25 m³/s. This small value is selected to prevent a large gradient in water level inside the channel especially at the locations where the water depth is small. If the water level drops drastically, this may cause instability in the model computation.

From day 10 to day 20, the total flow discharges increased from 50 m³/s to 160 m³/s depending on the sand discharge volume as can be seen in the second column of **Table 6-10**. This considerably high total flow discharge is selected to ensure that the sand bank can be fully removed from the channel. Nevertheless, from day 20 to day 30 the total flow discharges reduced slightly lower than the total flow discharges during day 10 to day 20. This is to ensure that the flow currents within the channel can be decreased. The reduced current may slow down the removal process of sand bank inside the channel. This removal process allows the residual sand to be completely flushed out from the channel, but at the same time prevents scouring that may occur at the channel's entrance.

Table 6-10: Combination of total flow discharges

Months	Sand discharges (m³)	Sand bank depths (m)	Q_1 (m³/s) 0-10 (days)	Q_2 (m³/s) 10-20 (days)		Q_3 (m³/s) 20-30 (days)	
April	45,325	3.69	25	120	150	80	100
May	27,031	2.20	25	60	90	25	45
June	63,363	5.16	25	180	210	150	170
July	53,621	4.37	25	140	170	120	140
August	56,251	4.58	25	160	190	130	150
September	37,325	3.04	25	80	110	60	80
October	20,206	1.65	25	50	60	25	20
November	5,174	0.42	25	30	30	5	10
December	38,813	3.24	25	100	130	50	70

*Area of sand bank = 12282.36 m² ; Blue= simulations without wave; Red = simulations with wave

Figure 6-22 shows the temporal evolution of the modelled sand bank in the channel along the cross-shore channel profile based on the combination of total flow discharge for the cases (a) without wave and (b) with wave for two different flow discharge rates. This result plot is based on the sand bank discharge of April 2011. It should be noted that in all cases the initial height between the sand bank crest level and initial water level is 1.0 m to allow sufficient variations in water level so that the numerical instability can be avoided.

For the case when the model is run without wave, result shows significant accretion of sand in front of the channel's entrance. In this particular case, the longshore current is quite minimal to revolutionise the accumulated sand in front of the channel's entrance. In contrast, for the case when the model is run with wave, the result shows less accumulation of sand in front of the channel's entrance. Much of the sand bypasses the Snapper Rock headland and is distributed to the neighbouring beaches due to the strong wave-induced longshore currents. The total flow discharges that are modelled with the inclusion of waves are used for the next morphological model runs.

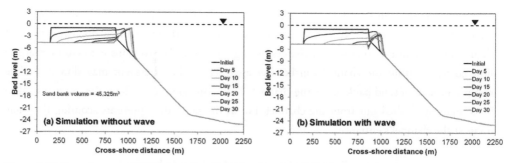

Figure 6-22 Cross-shore profile evolution of computed sand bank modelled by a model without wave and a model with wave. The total flow discharge rate is different for both simulations.

Sand bank discharge operations

We utilised the same initial bathymetry (i.e. March 2011) for each month starting from April 2011 to December 2011. Only the bed level (sand bank) in the artificial channel is updated depending on the total sand discharge volume as shown in **Table 6-10**. The reason why we keep using the same initial bathymetry for each month is because we only interested to observe the sand distribution patterns due to the effect of different sand discharge rates and different wave climates rather than analysing the morphological beach behaviour of a full cycle of a year. The study area comprises a complex coastal system and some considerations should be taken if the full morphological run is applied.

Figure 6-23 shows the modelled evolution of the morphological bed change patterns with and without the inclusion of sand bypassing system. For this particular case, model simulation was carried out from 1st April 2011 to 30th April 2011. In a case when the sand bypassing is considered, a total of 45,325 m³ of sand bank is initially placed on the artificial river bed channel. The bed was updated after 12 hours.

The discharged sand bank is obviously seen in **Figure 6-23** during the 10 simulation days. This is observed when the underwater bed contour lines start to migrate seawardly in front of the channel's entrance indicating an accumulation of sand. During the first 10 days, areas in front of the Rainbow bay beach are largely affected by the artificial sand bypassing system. Much of the sand from the Snapper Rock headland is deposited in front of the Rainbow bay beach. In a case of bypassing, a sandspit develops just in front of the Rainbow bay beach. In a case of no bypassing, the bed contour lines in front of the Rainbow bay beach evolve. Sandspit is observed, however its magnitude is relatively smaller than the magnitude of sandspit that is predicted by the model with the inclusion of sand bypassing. On day 20, the modelled sandspit grows bigger and slowly moves towards the Greenmount headland in a case of bypassing. On the other hand, the sandspit disappears in a case of no bypassing. Areas in front of the Snapper Rock headland have been eroded. By day 30, the underwater bed contour lines are relatively straight and areas in front of Snapper Rock headland keep on eroding. In the case of bypassing, the elongated sandspit seems to keep on moving and eventually bypasses the Greenmount headland after 30 simulation days.

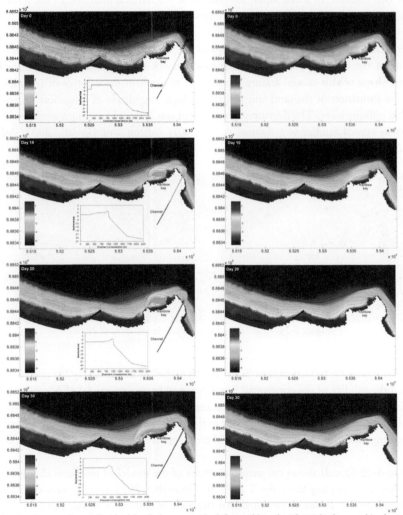

Figure 6-23. Evolution of modelled bed level changes with (left panels) and without (right panels) the inclusion of sand bypassing system. Independent insets in the left panels show the cross-shore evolution of sand bank in the channel. In all panels iso-contours (0.5m intervals) are contoured in the background.

The sand bypassing phenomenon observed in the model is relatively identical to the one observed in reality (see **Figure 6-24**). In **Figure 6-24**, it can be seen that a significant amount of sand is deposited in front of the Rainbow bay beach. The dotted black lines in **Figure 6-24** indicate the distribution pattern of sand. The shape of this sand pattern more or less represents the shape of the sandspit that was predicted by the model.

The sand bank discharge operation has contributed to the morphological evolution of the nearshore bed changes. In a case when the sand bank discharge is included in the model, there is a formation of a sandspit in front of the Rainbow bay. Comparing the modelled bed change pattern (see **last panel of Figure 23**) with an aerial photograph (see **Figure 24**), the morphological feature of the sandspit appears in both photos. Despite the fact that the magnitude of the sandspit predicted by the model is not exactly similar to the one observe in reality, the shape of this coastal feature is sufficient to confirm the validity of the process based model. The formation of the sand spit is due to a high magnitude of longshore currents caused by a highly oblique wave. The development of a sand spit induced by a high-angle wave has been presented in several studies (e.g Ashton et al., 2001; Ashton and Giosan, 2007; Ashton and Murray,2006).

Figure 6-24. Coalangatta bay condition based on April 2011.

The sand bypassing process as seen in the model as well as in the reality for this particular case has proved the validity of the conceptual headland sand bypassing model of Short and Masselink (1999). In the model of Short, the sand bypassing occurred when the sand wave started to move around the headland and creating a sandspit at the leeside of the headland. Detailed processes of sand bypassing based on the conceptual model are given in **Section 2.3.1** in **Chapter 2**.

Figure 6-25 (b,c,d) shows the snapshot results of the modelled bed level changes at the end of each month beginning from April 2011 to December 2011 for the non-continuous sand bypassing operations. It should be noted that the wave is coming from the north-northeast from December to April 2011. Alternatively, from May to November 2011 the wave is coming from the south-southeast directions. Figures below each modelled bed level changes panel are the offshore wave rose plots for each associated month. The initial model bathymetry is plotted in **Figure 6-25 (a).**

Referring to all panels in **Figure 6-25 (b,c,d)**, we can see that the modelled nearshore bed contour lines especially in front of the Rainbow bay beach show different behavioural patterns for each month. In a case when the sand bank discharge is less, the model result shows disappearance of the sandspit. In a case of large sand bank discharges, the pulses of sand manifest themselves as large sandwaves moving around Greenmount headland and finally attaching themselves to Coolangatta beach. As can be seen in **Figure 6-25(b)-panel June 2011** and **Figure 6-25(c)-panels August 2011**, the tail of the sandspit that is formed by the sand waves is observed at the edge of Greenmount headland which is evidence of the massive sand delivery updrift of the Rainbow bay coast. Qualitatively, the modelled sand spit as observed in both figures is identical to the actual sand spit that is observed from the aerial images as showed in **Figure 6-26**. In **Figure 6-26**, the actual sandspit is clearly seen attaching itself to the edge of the Greenmount headland. For a record, the pumped sand volume during August 2002 and June 2007 is 117,842 m³ and 60,129 m³ respectively, which is considerably large.

In **Figure 6-27**, we show the differences in bed level after 30 days of simulation. Obviously, the sand bank discharge operations have strong effects, both negative and positive. In all cases, area in front of the Snapper Rock headland is significantly eroded due to high longshore velocities. The high velocities are contributed by a combination of the wave induced velocities and river discharge velocities. The dispersion of sand is noticeable when the deposition of sand is observed in front of the Greenmount headland.

Exceptional is given to the model result of December 2011, where there is significant change in bed level contours at the end of simulation time. As can be seen in the last panel of **Figure 6-25(d)**, the bed contour lines smear out and the area immediately in front of Snapper Rock is scoured as a result of strong currents which are caused by high wave energy. This high wave energy with significant wave height over 4.0 m arriving from the East-Northeast is due to ex-tropical cyclone Fina. This event occurred at the end of December 2011. The sedimentation/erosion plot of **Figure 6-27 (panel December 2011)** indicates the whole stretches of the coast starting from the Snapper Rock headland to Kirra beach are significantly affected by the cyclone Fina.

Effect of seasonal wave climates

Bi-seasonal wave climates may influence the development of the sand spit and thus affect sand distribution patterns around the southern Gold Coast beaches. The southern Gold Coast beaches are influenced by the changes of wave climate from the northeast and southwest directions. Incoming waves, particularly wave directions, seem to play an important role in the formation and movement of the sand spit. In a case of large sand bank discharge supplies, model result shows that easterly waves and currents cause the elongated spit to bypass the Greenmount headland (see **Figure 25(b)-panel April 2011**). On the other hand, the model result shows that the persistent south easterly waves and currents move large amount of sand around the Snapper Rock headland. This sand temporarily builds up at Rainbow bay, creating a sandpit and wider beach develops (see **Figure 25-panel(c) July 2011**).

The dynamic sand distribution patterns caused by the changing of seasonal wave climates and due to the permanent sand bypassing system can be seen in **Figure 28**. The north easterly waves transport sand beyond the Greenmount headland while, large amount of sand is deposited in front of the Rainbow bay which is protected from the southerly waves by the Snapper Rocks headland.

The influence of wave climates on large sand distribution patterns around headlands has been investigated in previous studies. The sand bypassing of the Slade Point headland and the Eimeo headland, for instance tends to be episodic in response to the variability of the wave climates (Environmental Protection Agency, 2005). The bypassing of the Slade Point headland is strongly associated with storm wave events such as cyclone swell refracting around the headland and the northeasterly storms, and results in a series of sand shoals forming in the lee of the headland. High transport rates caused by the events with a mainly southerly component tend to move large quantities of sand around the Slade Point headland, which then accumulate in the sheltered zone as a series of distinct banks. Likewise, at the Eimeo headland a proportion of sand is transported along the western shore of the headland towards the Eimeo Creek's entrance. This sand bypassing is episodic in nature and the shoreline often shows significant fluctuations as slugs of sand move through.

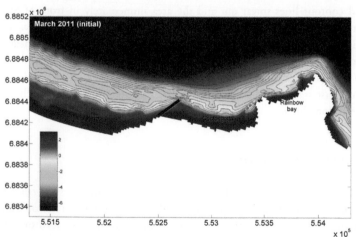

Figure 6-25 (a). Initial model bathymetry setup based on March 2011. Iso-contours (0.5m intervals) are contoured in the background.

6. Impact of a permanent sand bypassing on the natural sand distribution patterns, southern Gold Coast beaches, Australia

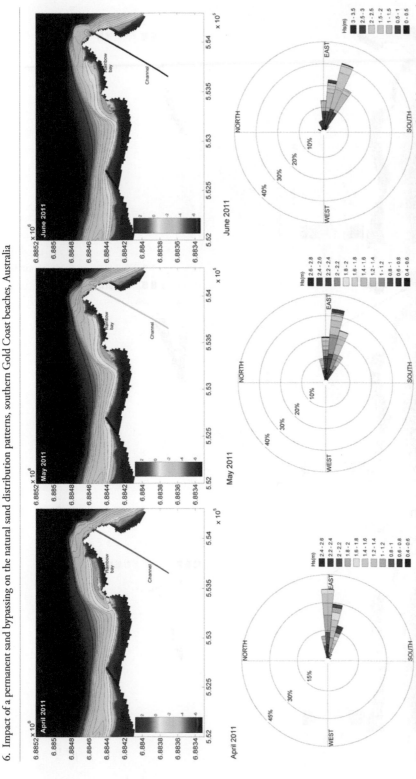

Figure 6-25(b). Snapshots of the modelled morphological bed contour changes at the end of each month (April, May and June) with the inclusion of sand bypassing system and associated wave roses. In upper panels iso-contours (0.5m intervals) are contoured in the background.

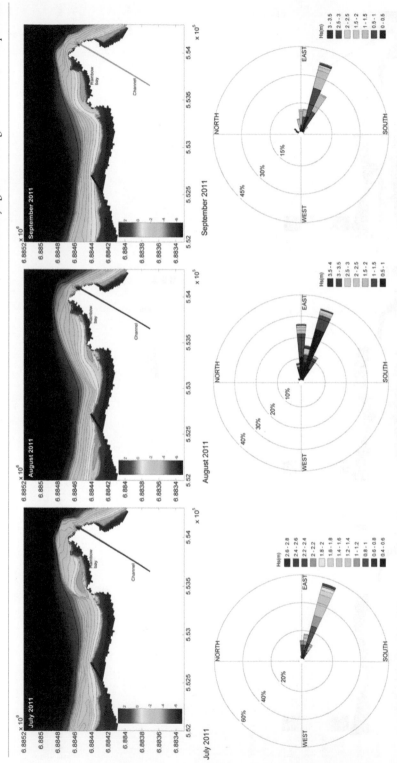

Figure 6-25(c). Snapshots of the modelled morphological bed contour changes at the end of each month (July, August, September) with the inclusion of sand bypassing system and associated wave roses. In upper panels iso-contours (0.5m intervals) are contoured in the background.

6. Impact of a permanent sand bypassing on the natural sand distribution patterns, southern Gold Coast beaches, Australia

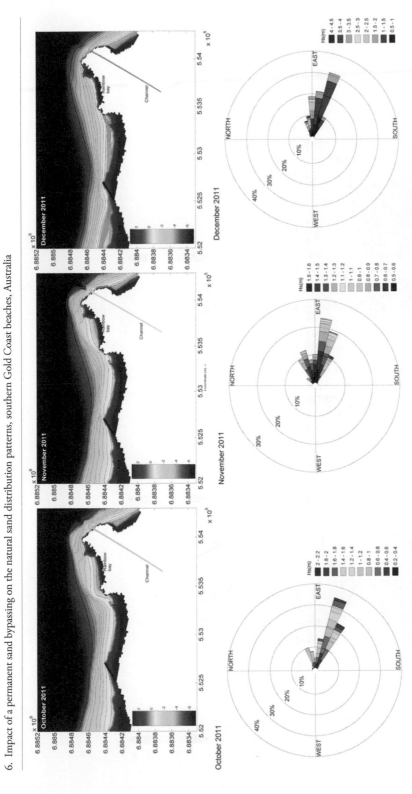

Figure 6-25 (d) Snapshots of the modelled morphological bed contour changes at the end of each month (October, November, December) with the inclusion of sand bypassing system and associated wave roses. In upper panels iso-contours (0.5m intervals) are contoured in the background.

Figure 6-26. The formation of an elongated spit due to a large sand supply discharged from the sand pumping outlet (Worley Parsons, 2009).

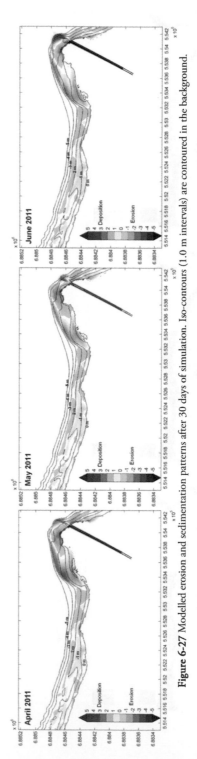

Figure 6-27 Modelled erosion and sedimentation patterns after 30 days of simulation. Iso-contours (1.0 m intervals) are contoured in the background.

6. Impact of a permanent sand bypassing on the natural sand distribution patterns, southern Gold Coast beaches, Australia

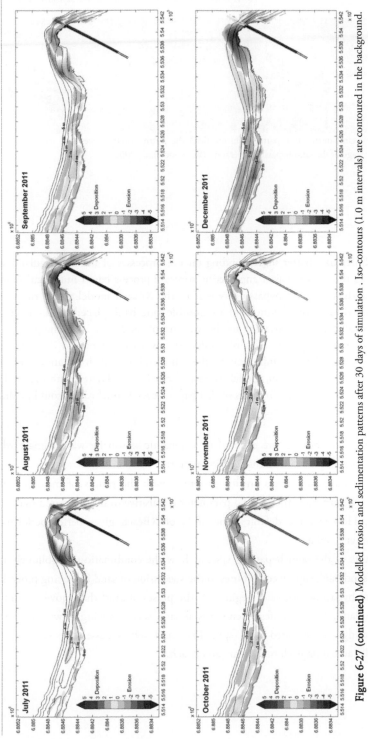

Figure 6-27 (continued) Modelled rrosion and sedimentation patterns after 30 days of simulation . Iso-contours (1.0 m intervals) are contoured in the background.

Figure 6-28. The natural sand distribution patterns during (left) the north easterly waves and (right) the south easterly waves in the early stage of the sand bypassing operation (Boswood et al., 2005).

6.6 Conclusions

This study contributes to a better understanding on the processes of natural headland sand bypassing by extending the use of coastal engineering tools i.e process based numerical models. In this study, we introduced a sand bypassing system in the XBeach model where the system comprises of a long fixed bed river channel and an erodible sand bank placed on the river bed. The volume of the sand bank is equivalent to the volume of sand that is discharged by the mechanical sand bypassing pumps. A series of model tests was carried out to determine the suitable total flow discharges that can fully remove the sand bank out of the channels. Prior to morphodynamic modelling investigation, models were calibrated and validated with the available field data measurements. Based on the findings of this paper, several conclusions could be drawn and listed as follows:

1. The southern Gold Coast beaches are exposed to a dynamic seasonally wave climate which significantly affects the longshore sand distribution patterns.
2. Results of modelled wave parameters are in agreement with the results of measured wave parameters. The only short wave friction (f_w) parameter is sensitive to the measured significant wave height data. However, the default value of f_w of XBeach gives better performance compared to the other f_w values.
3. The modelled results of the sand bypassing system show the combination of obliquely high waves and greater sand bank supply contributes to the succession of sand bypassing processes.
4. The modelled sand bypassing processes begin with the pulses of sand that move around the Snapper Rock headland. They manifest themselves as sand waves moving slowly around the Rainbow bay creating an elongated sand spit. The sand spit bypasses the Greenmound headland and finally attaching itself to Coolangatta beach.

5. The modelled sand bypassing process is identical to the conceptual headland sand bypassing process of Short and Masselink (1999). The conceptual headland sand bypassing model is valid only if there is enough supply of sand updrift of the structure.

6. Seasonal wave climates contribute to the dynamic sand distribution patterns around the southern Gold coast beaches. South easterly waves temporarily retain the sand in front of the Rainbow bay. The beach continues to increase in width until sand at the opposite end of the Rainbow bay begins to flow around the Greenmount headland into Coolangatta bay. While, the north easterly waves and currents bring the sand to bypass the Greenmount headland.

7. In the case of bypassing, the high delivery of sand is able to assist in restoring the head of the Snapper Rocks sand bank, and further grooming of the underwater nearshore bed contours.

This numerical modelling study has been conducted in a complex coastal environment. There are few limitations that needed to be addressed so that recommendations can be offered for further research improvements.

In this study, the flow discharge is determined based on the trial and error method, which is time consuming and requires high computational efforts. In the present modelling study, the river discharge option in Xbeach allows the uni-directional inflow computation only, without sediment transport. In addition, the inflow boundary must be located at the land (wall) boundary. The model could be improved by introducing a sediment concentration at the inflow boundary in the model formulation codes. If the sediment concentration is introduced at the inflow boundary, therefore the input sediment discharge can be easily computed. This of course, saves lots of computational time. In addition, the inflow boundary should be placed at the edge of the coastline i.e somewhere in the middle domain instead of at the land boundary. This way, a long river fixed bed channel and an erodible sand bank is not necessarily needed.

In this modelling study, the presence of the Tweed River is neglected due to the fact that sediment that is discharged by the river is considered insignificant. As the study area is characterised as wave-dominated, the role of tide is simply ignored. The presence of ebb and flood tides however may also influence the outgoing and incoming sediment transport patterns of the Tweed River which are caused by the ebb- and flood-induced currents. Incorporating Tweed River channel and tide is important and may give different model results.

The morphological model developed in this study is restricted to a non-continuous sand bypassing operation which means the nearshore bathymetry for the next morphological model run is not updated. This way, the morphological beach behaviour of a full cycle year cannot be captured. In the future study, a full year morphological run can be done but some model considerations should be taken into account. Sand removal at the Tweed River jetty, for an instance should be performed to prevent a huge accumulation of sand behind the jetty which is caused by the strong wave-induced longshore transport.

References

Acworth, C., and Lawson, S. (2012). The Tweed River entrance sand bypassing project, ten years of managing operations in a highly variable coastal system, *Proceeding of the 20ʰ NSW Coastal Conference 2012 Tweed Head (2012)*, pp 1-23

Ashton, A.D., and Murray, A.B. (2006). High-angle wave instability and emergent shoreline shapes: 2. Wave climate analysis and comparison to nature. Journal of Geophysical Research. 111, F01012, DOI:10.1029/2005JF000423.

Ashton, A.D., Murray, A.B., and Arnault, O. (2001). Formation of coastline features by large-scale instabilities induced by high-angle waves. *Nature*. Vol. 414(): 296-299.

Ashton, A.D., and Giosan, L. (2007). Investigating plan-view assymetry in wave-influenced deltas. In: *5ʰ IAHR Symposium on River, Coastal and Estuaries Morphodynamics*, International Association of Hydraulics Researchers, Enschede. pp 681-688.

Boswood, P.K., and Murray R.J. (2001). Worldwide sand bypassing system: Data report. Conservation Technical Report No.15, Queensland Australia. ISSN 1037-4701 August 2001.

Boswood, P., Victory, S., Lawson, S. (2001). Placement strategy and monitoring of the Tweed River Entrance Sand Bypassing Project nourishment work. *Proc. Coasts and Ports Conference 2001*, pp. 253–258.

Boswood, P.K., Voisey, C.J., Victory S.J., Robinson D.A., Dyson A.R., and Lawson. S.R. (2005). Beach Response to Tweed River Entrance Sand Bypassing Operations. *The International Conference of Coasts & Ports 2005*, 20-23 September 2005.

Castelle, B., Lazarow, N., and Tomlinson, R. (2006). Impact of beach nourishment on Coolangatta bay morphology over the period 1995-2005. *Proceeding of 15ʰ NSW Coastal Conference*, pp.1-12

Castelle, B., Bourget, J., Molnar, N., Strauss, D., Deschamps, S., and Tomlinson, R. (2007). Dynamics of a wave-dominated tidal inlet and influence on adjacent beaches, Currumbin Creek, Gold Coast, Australia. *Coastal Engineering*, 54(2007): 77-90.

Castelle, B., Bourget, Turner, I.L., Bertin, X., and Tomlinson, R. (2009). Beach nourishments at Coolangatta Bay over the period 1987-2005: Impacts and lessons. *Coastal Engineering* (2009): 1-11.

Colleter, G., Cummings, P., Aguilar, P., Walters, R., and Boswood, P. (2001). Monitoring of Tweed River Entrance dredging and nourishment activities. *Proc. Coasts and Ports Conference 2001*, pp. 259–264.

Cheung, K.F., Gerritsen, F., and Cleveringa, J. (2007). Moprhodynamics and sand bypassing at Ameland Inlet, the Netherlands. *Journal of Coastal Research*, 23(1): 290-299.

Dyson, A., Victory, S., and Connor, T. (2001). Sand bypassing the Tweed River Entrance: an overview. *The 15ʰ Australian Coastal and Ocean Engineering Conference, the 8th Australian Port and Harbour Conference 25-28 September 2001*, Queensland Australia.

Environmental Protection Agency. (2005). Mackay Coast Study. The State of Queensland Environmental Protection Agency, 113p

FitzGerald, D.M., Kraus, N.C., and Hands, E.B. (2001). Natural mechanisms of sediment bypassing at tidal inlets, US Army Corps of Engineers, USA. pp.1-10.

FitzGerald, D.M, and Pendleton, E. (2002). Inlet formation and evolution of the sediment bypassing system: New Inlet, Cape Cod, Massachusetts. *Journal of Coastal Research*,(): 290-299.

Hobbs, J.E., and Lawson, S.W. (1982). The tropical cyclone threat to the Queensland Gold Coast. *Applied Geography*, 2(): 207–219.

Legates, D. R., and McCabe, G.J. (1999). Evaluating the use of "goodness-of-fit" measures in hydrologic and hydro-climatic model validation. *Water Resources Res.*, 35(1): 233-241.

Keshtpoor, M., Puleo, J.A., Gebert, J., and Plant, N.G. (2013). Beach response to a fixed sand bypassing system. *Coastal Engineering* (73): 28–42.

Moriasi, D.N., Arnold, J.G., Van Liew, M.W., Bingner, R.L., Harmel, R.D., and Veith, T.L. (2007). Model evaluation guidelines for systematic quantification of accuracy in watershed simulations. *American Society of Agricultural and Biological Engineers*, Vol. 50(3): 885–900.

6. Impact of a permanent sand bypassing on the natural sand distribution patterns, southern Gold Coast beaches, Australia

---

Roelvink, J.A. (1993). Surf beat and its effect on cross-shore profiles. PhD Thesis, Technical University of Delft, the Netherlands, 150 pp.

Short, A. D. and Masselink, G. (1999). Embayed and structurally controlled beaches, In: Short, A.D.(ed)., *Handbooks of Beach and Shoreface Hydrodynamics*. Chicester: John Wiley & Sons. pp. 230-249

Stuart, G., and Lewis, J. (2011). Gold Coast shoreline management plan, Field measurement and data collection: Technical report June 2011, DHI Water and Environment Pty Ltd, Southport Queensland. 34 pp

Strauss, D., Burston, J., Girondel, T., and Tomlinson, R. (2013). Multi-decadal analysis of profile response to permanent bypassing. *Coastal Dynamics*, pp 1547-1558

Turner, I.L., Aarninkhof, S.G.J., and Holman, R.A. (2006). Coastal imaging applications and research in Australia. *Journal of Coastal Research* 22 (1): 37–48.

Tomlinson, R., Lazarow, N., and Castelle, B. (2007). Kirra wave study: Report No.59, Griffith Centre for Coastal Management, Southport, Queensland Australia. 28pp

van Dongeren, A., Lowe, R., Pomeroy, A., Duong, T., Roelvink, D., Symonds, G., and Ranasinghe, R. (2012). Modelling infragravity waves and currents across a fringing coral reef. *Coastal Engineering Proceedings*, 1(33), currents. 29. DOI :http://dx.doi.org/10.9753/icce.v33.currents.29.

Worley Parson. (2009). Kirra groyne effects study: Report No. 301001-00826 – 001, Worley and Parsons and Land and Property Management Authority, Australian, 59p.

CHAPTER 7

Summary and conclusions

This chapter summarises the overall works presented in the thesis. Research answers in response to each research question are addressed. The practical implications of the findings and prospects for further researches are offered.

7.1 Conclusions

The thesis in general explores the mechanisms and processes of sand bypassing. The thesis combines collective efforts of numerical modelling works in an artificial coastal environment and a non-artificial coastal environment. The numerical process based model of XBeach is mainly used to perform all scenario tasks in order to answer some research questions that were formulated in the **Chapter 1** of this study.

In this study, sand bypassing rates and shoreline evolution near coastal (groyne) structure are investigated by comparing analytical solutions and XBeach numerical modelling. The scheme of headland sediment bypassing has been proven through the numerical modelling works.

The behavioural patterns of hydrodynamic rip currents and morphological shoal and/or bar patterns predicted by the XBeach model are in agreement with the empirical models. The hypothesis that sediment in deeply embayed beaches could be transported outside the surf zone by the surf zone currents is valid.

Results obtained from the field investigation study on a small embayed beach has shown that seasonal wave climates contribute to the onshore-offshore sand movements. Sand leakage around a small headland is caused by the beach rotation process.

The combination of the sand discharge operation and the natural action of waves has proven to contribute to an excessive delivery of sand to the neighbouring beaches. The development of sandspit indicates a positive role of an additional sand supply which is successfully delivered by the sand bank discharge operations.

The natural sand bypassing processes can be completed in a matter of days to years. In this study, both short and long term model simulations were conducted. A long term model simulation of sand bypassing running with the XBeach program is unique and has not been popularly used by many researchers due to its heavy computational times. A multi-core-processors approach (MPI) used to execute the XBeach program helps to minimise the long computational times. For computational efficiency, models were run on Linux based operating system computers. Besides, the innovation of instationary (wave groups) wave model of XBeach, by means of incorporating the contribution of infragravity waves applied in all model cases presented in this study bridging the gaps between the previous studies and this present study. In this study a sand bypassing system was simulated through the use of an artificial river inflow. In future work this should be adapted to a number of point sources and sinks with specified flow and sediment discharges.

7.2   Responses to research questions

In this section, the research questions as presented in the **Section 1.3** are addressed. The main research questions are re-written and responses to each research question are given as follows:

1. What are the mechanisms of sand bypassing and how the processes of sand bypassing work?

In **Chapter 2**, the mechanisms of sand bypassing were identified and the processes of sand bypassing were explained. There are two main mechanisms of sand bypassing which were classified into wave-driven alongshore sand bypassing and wave-driven cross-shore sand bypassing. For the wave-driven alongshore sand bypassing mechanism, oblique waves are necessitated to drive the littoral sand along the shore. The presence of coastal structures like groynes or headlands on the coast blocks the movement of sand along the shore. As a result, sand is trapped on the up-drift side of the structure. The volume of trapped sand on the up-drift structure increases over the times and eventually reach up to the tip of the structure. Likely, sand continuously moves around the structure's tip, attaching itself to the down-drift side of the structure and finally merges to the down-drift coast. The bypassing of sand most likely occurs under the water as a subaqueous sand layer. For the wave-driven cross-shore sand bypassing mechanism, sand must be first moved cross-shore, out past the surfzone. The cross-shore movement is most likely occurs due to rip currents. Once the sand has moved offshore, it must be advected and remain in suspension. A small wave height obliquely approaching the shoreline could initiate alongshore currents that would sweep the sand and deposit it in the nearshore, feeding the downcoast beaches. It appears that headland rips can transport sand beyond the headland boundary and sand may be deposited farther away to the offshore.

2. What are the main driving forces (factors) that may contribute to the sand bypassing process?

There are the several main driving forces and factors that contribute to the succession of sand bypassing processes. In **Chapter 3**, we have presented the effect of increasing wave parameters (wave height and wave angle) and the variation of sediment grain sizes on the shoreline evolution patterns, sand bypassing processes and sand bypassing volumes. The numerical model results showed that increasing the wave angle, wave height and sediment grain size leads to the increment of sand bypassing volume. Nevertheless, in all cases the processes of sand bypassing around a groyne structure remain unchanged. The inclusion of infragravity wave in the models showed that the computed model results of shoreline evolution patterns mimicked the shoreline patterns as observed in reality. In **Chapter 4**, the model results showed that increasing the wave height had limited the development of surf zone currents in the middle of embayment basins by creating a large seaward flowing rip current. This rip current enhances the transportation of littoral sand outside the surf zone. The model results based on the effect of wave directional spreading had shown to contribute to the generation of surf zone rip currents and shoal bars. The presence of the headland rips in the model is favoured by the large wave directional spreading. In the case when virtual drifters are employed on an embayed beach, the percentage of drifter exits increases when the infragravity waves are applied. Headland rips are responsible to transport the sand outside the embayment basin. In different sizes of embayment basin, a small increase in wave angle contributes to the sand bypassing around headland. Based on the results obtained from the field observation analyses (in **Chapter 5**), wave seasonality (variations in wave heights and bi-directional waves) magnifies the onshore-offshore sand movement. Beach rotation which is caused by the effect of wave seasonality contributes to the sediment leakage around the small headland. The numerical model results presented in **Chapter 6** showed that a combination of sand pumping discharges and uni-directionally wave contributes to the succession of sand bypassing processes around small headlands and the groyne structure.

3. To what extent does wave seasonality affect the sand bypassing process ?

Wave seasonality is linked to the variation of wave heights and wave directions. In **Chapter 5**, the results of temporal and spatial cross-shore sand distribution patterns based on the beach profile surveys indicated that the dominant coastal sediment transport process is through the cross-shore sand movement. During the calm season, low energy waves push the sand back to the beach while during the storm season high energy waves bring the sand out to the sea. The seasonal wave variations (bi-directional waves) caused the beach to rotate and this rotation had contributed to sand leakage around a small headland (Tanjung Tembeling headland). Since the headland is located in a water depth shallower than the closure depth, sand could easily bypass the headland structure. In this particular case study, the results showed that both the cross-shore and alongshore sand bypassing processes are mutually involved.

4. To what extent does the permanent sand bypassing system contribute to the natural sand distribution patterns on a complex coastal environment?

Permanent sand bypassing system is one of the effective methods to deliver some amount of sand from an updrift coast to a downdrfit coast. In **Chapter 6**, the impact of permanent sand bypassing on the natural sand distribution patterns around natural headlands and a groyne structure was investigated. In the field site, sand is mechanically pumped from the updrift side of the Tweed River jetty and is discharged back to the sea within few minutes. The pumped sand is delivered by underground pipes that discharges the sand to several fixed discharge outlets. In the model, a long river channel is created and an erodible sand bank is build and placed on the channel's bed. Given a sufficient flow discharge at the inlet discharge boundary, the high flow currents will push the sand bank out of the channel to the sea. Sand will eventually accumulate in front of the channel's entrance and it is expected to be naturally distributed by the natural action of waves. The natural distribution of sand is done by the wave refraction process. Incoming high energy swell waves are generally refracted around the Snapper Rock headland enhancing a strong flow current around the headland that eventually distributes the sand to the neighbouring beaches. Without the permanent sand bypassing system, the energetic swells may bring the sediment away from the coast. As a consequence, beaches are continuously eroded. In **Chapter 6**, the numerical modelling results show that the sand bypassing process begins when the sand first moves around the Snapper Rock headland. They manifest themselves as sandwaves, moving around the Greenmount and creating an elongated sandspit. This sandspit grows bigger bypasses the Greenmount headland and finally attaching itself to Coolangatta beach.

5. Are the results obtained from the process based model comparable to the results obtained in the empirical and analytical models ?

In this study, both analytical and empirical models are used and the results are compared to the results obtained from the process based model. In **Chapter 3**, the analytical shoreline evolution and sand bypassing models of Pelnard were applied. The results of shoreline evolution patterns and sand bypassing volumes obtained from the analytical Pelnard model were compared with the results of shoreline evolution patterns and sand bypassing volumes obtained from the process based XBeach model. Both analytical and process based models were in a good agreement. Nevertheless, differences in model results were observed between the process based and analytical models. These result's discrepancies might be due to the prior assumptions made in the analytical model. In **Chapter 4**, the use of empirical models (embayment scaling parameter models) to characterise the surf zone current circulations and embayed beach morphology has been proven to work well with the process based XBeach model. The Xbeach model can reproduce the morphological shoal and bar patterns and associated surf zone rip currents as predicted by the empirical models.

6. Do the sand bypassing processes simulated by the process based model represent the sand bypassing scenarios in reality ?

The use of two-dimensional depth-averaged (2DH) process based model in this study is relevant to investigate the horizontal patterns of current flow, sediment transport paths and morphological bed changes. A combination of river discharges and an erodible sand bank structure built in a fixed river channel can be represented as an artificial sand bypassing system. The natural sand bypassing process demonstrated by the model can represent the natural sand bypassing process as in reality (refer to **Chapter 6**).

7.3 Practical implications of the findings and prospects for further researches

The outcomes of the present study demonstrate that understanding of the mechanisms and processes of headland sand bypassing is important as a base guideline to analyse the sedimentary budget for future coastal zone management in both artificial and non-artificial coastal environments.

Based on the findings of this study, both alongshore and cross-shore sand bypassing mechanisms require a mobilization of sand from one place to another place. The role of wave heights, wave angles, and sediment grain sizes are proven to contribute to the succession of sand bypassing processes. The determination of infilling time (t_{fill}), a period when the sand begin to bypass a groyne structure is important in defining proper mitigation measures to minimise localised erosion at the downdrift beaches.

In a case when the coasts are characterised by nearshore sand bars and are restricted by the presence of coastal structures (headland impacts), the formation of surfzone rip currents (headland rips, beach rips, megarips) prevails. These surf zone rip currents are found responsible to carry sand out from the beach to offshore. Comprehending the morphodynamic characteristics of surfzone rip currents is essentially important to further understand the long term stability of embayed coasts.

From the results of field observation study, it is learned that wave seasonality magnifies the cross-shore transport process in a small embayed beach. The alternating behaviour of nearshore sand bars results in a classical coastline pattern. The bi-seasonal wave climates cause the beach to rotate at one side of the beach. As a result, sand leakage around the headland may be expected. Indeed, beach nourishment has increased the width of the beach. The idea of nourishing the beach and the foreshore areas with sand may temporally hold the sand on the beach for a short term duration only. Alternative measures should be considered to protect the beach from the continuous erosion.

The outcomes of the final model investigation which are related to an artificial sand bypassing system provide an overview of natural sand distribution patterns due to combination between the natural wave action and the sand bank discharge operation. The natural sand bypassing is achieved via the straight and wide nearshore sand bar in addition to the alongshore shore beach break. The formation of the nearshore bar along the southern Gold Coast beaches causes the high energy of waves to break on it. As a consequent, the wave energy from the breaker line towards the beach is less dissipated thereby minimizing erosion rate at the beach. The implementation of an innovative artificial sand bypassing system in the XBeach model helps to understand the mechanism and processes of sand bypassing in a complex coastal environment. Indeed, the artificial sand bypassing system has helped to naturally nourish the neighbouring beaches by supplying a sufficient volume of sand.

Further research is required to investigate the role of tide on the natural sand bypassing processes around natural headlands or coastal structures. Details numerical investigation on the

181

formation of megarip currents on an embayed beach should be carried out as this factor (megarips) may significantly contribute to the cross-shore sediment exchange processes. On the other hand, a comprehensive field measurement should be carried out both for Malaysian and Australian case studies. The outputs from the field studies can be used to further validate the results of numerical model. Last but not least, mutual co-operations from respected stakeholders (e.g. government agencies, research institutions, universities) are highly requested for a better improvement of the present study.

About the author

Mohd Shahrizal bin Ab Razak was born in Pekan Pahang, Malaysia. He started his undergraduate study in Kolej Universiti Tun Hussein Onn Malaysia in 2001 and obtained his Diploma Degree in Civil Engineering Technology in 2004. In 2005, he was offered a scholarship under a Young Lecturer Polytechnics scheme by the Ministry of Education Malaysia to pursue his bachelor degree at the Universiti Tun Hussein Onn Malaysia, Batu Pahat Johor. He earned his Bachelor Degree in Civil Engineering (Hons.) in 2007. Immediately upon the completion of his bachelor degree, he continued his study in Master Degree in Technical and Vocational Education at the same university where he enrolled his bachelor degree. During his master study, he was offered an academic position at the Civil Engineering Department, Universiti Putra Malaysia and was recruited as a Tutor. In August 2008, he was granted a scholarship (IPTA Academic Training scheme) from the Ministry of Higher Education Malaysia to further his postgraduate engineering study. He gained his Master Degree in Hydraulics and Hydrology from the Universiti Teknologi Malaysia in 2009. In September 2010, he got admitted to a joint Ph.D programme between UNESCO-IHE Institute for Water Education and Delft University of Technology, the Netherlands to pursue his study in coastal morphodynamics modelling. His Ph.D study was fully sponsored by the Ministry of Higher Education Malaysia under the Bumiputra Academic Training scheme scholarship. He has published several articles in scientific journals and has presented his works in both national and international conferences.

List of publications:

Ab Razak, M.S., Dastgheib, A., Roelvink, D., Strauss, D. (2015). Headland sand bypassing: A proof of concept. *Journal of Coastal Research.* (submitted)

Ab Razak, M.S., Dastgheib, A., Roelvink, D., Suryadi, F.X. (2015). A process-based model of a permanent sand bypassing system. *Journal of Coastal Research.* (under review)

Ab Razak, M.S., Dastgheib, A., Suryadi, F.X., Roelvink, D. (2014). Headland structural impacts on the surf zone current circulations, *Journal of Coastal Research*, SI 70: 65-71.

Ab Razak, M.S., Roelvink, D., and Reyns, J. (2013). Beach response due to sand nourishment on the east coast on Malaysia. *Proceeding of the ICE-Maritime Engineering Journal*, 166(4): 151-174.

Ab Razak, M.S., Dastgheib, A., Roelvink, D. (2013). Sand bypassing and shoreline evolution near coastal structure, comparing analytical solution and XBeach numerical modelling, *Journal of Coastal Research*, SI 65: 2083-2088.

Ab Razak, M.S., Dastgheib, A., Roelvink, D. (2013). An investigation of sand bypassing parameters and bypassing solution of a groyne structure, *Proceeding of the International Conference of Coast and Port 2013*, Sydney, Australia.

Ab Razak, M.S., Dastgheib, A., and Roelvink, D. (2014). Morphodynamic investigation of embayed beaches through the impact of structural headlands, *Abstract in : NCK Days 2014.* Delft, the Netherlands.

Ab Razak, M.S., and Roelvink, D. (2011). Determination of equilibrium stages of headland bay beaches: A preliminary study on the east coast of Malaysia. *Proceeding of the International Conference of River, Coastal and Estuarine Morphodynamics (RCEM2011),* 6-8 September 2011,Beijing China.

Others:

Suryadi, F.X., Kurniawati, M., **Ab Razak, M.S.,** Marpaung, F.M., Kurniawan, B. (2015). River Restoration in DKI Jakarta, Indonesia. A Case Study of Ciliwung River. The 36th IAHR World Congress, 28 June - 3 July 2015, Delft-The Hague, the Netherlands. *Paper was accepted for oral presentation.*

Ab Razak, M.S., Sembiring, L., Van der Wegen, M. (2014) Preparedness. Book of Abstracts NCK-Days 2014, 82p.

Suryadi, F.X., Baedlowi, N., **Ab Razak, M.S.,** Kalmah (2014). Hydraulic performance of urban polder water management and flood protection systems in Jakarta. *Proceeding of the 13th International Conference on Urban Drainage,* 7-12 September 2014, Sarawak, Malaysia.

Badronnisa Y, Goi., **Mohd Shahrizal A.R.,** Abdul Halim G.,Thamer M.A,. (2010). Assessment of Sediment Transport Formula: A Case Study of Muda River. *Proceeding of the World Engineering Congress (WEC),* 2-4 August. Kuching, Malaysia.

Ab Razak, M.S., and Ismail, H. (2010). Beach Response Due to the Pressure Equalization Modules (PEM) System. *Proceeding of the National Conference of Hydrology and Environment (HIDRAS),* Universiti Tun Hussein Onn Malaysia, 23-24 June Batu Pahat Johor, Malaysia.

Kaprawi, N., Razzaly, W., Apandi, S.M., **Ab Razak, M.S.** (2007). Hubungan Antara Industri dan Institusi Pendidikan Tinggi Berorientasikan Teknikal. *Proceeding of the National Conference of Continuing Technical Education & Training (CCTET),* 11-23 September, Batu Pahat Johor, Malaysia.

Acknowledgements

Living in a foreign country for quite a long time has taught me to be an independent person. Emotion, loneliness, sadness, happiness, depression, fear are all sort of feelings that I had experienced during my Ph.D study in the Netherlands. Nevertheless, I am very grateful to have shared all those kinds of feelings with lots of great people during my stay in Delft and to that extent I would like to dedicate my deepest gratitude to all of you.

First of all, I would like to acknowledge my sponsor the Ministry of Education Malaysia and my employee Universiti Putra Malaysia for providing a full financial support for this four years Ph.D programme. A special appreciation is given to the Dutch national e-infrastructure with the support of SURF Foundation for providing a high performance cloud facility to support my research works.

Next, my great acknowledgement is extended to my promoter Prof.ir. dr. Dano Roelvink for his willingness to be my supervisor. It was a great experience working with you for the last four and a half years. Although our meeting time always short, the outputs and feedbacks were valuable and beneficial for the improvement of my Ph.D works. I would like to convey my thanks to Dr. Ali Dastgheib, my mentor, for guiding me to conceptualise the idea to structure the content of this thesis in such a logical way and eventually making the thesis readable. Your word by word explanation inspires me to be a good teacher in future. My acknowledgements are further conveyed to all committee members for their time in reviewing the content of this thesis. My acknowledgments are also dedicated to colleagues and staff members in the Coastal System & Engineering and Port development core group: Johan, Fernanda, Guo, Wan, Duong, Dr. Mick, Prof. Ranasinghe, Gerald, Dr. Poonam, Maurits, Hao, Liqin Zuo, Abdi and Duoc.

Studying in an international water education institute, I have made lots of international friends. Sharing ideas and thoughts about cultures, foods, religions and sometimes political issues are part of the conversation topics during our leisure times. Thank you all for being part of my family in Delft. Indeed, I have a special acknowledgment to my closest colleagues and dearest friends.

I am very much indebted to Isnaeni and Leo for their kindness of sharing ideas, mutual academic discussions and technical matters related to computer programming. Your endless support makes my dream come true. I also wanted to say thank you to my ex-front neighbour in the office, Yos, who always listens and cares about my problems whenever I need his help. You are like my brother and I really hope that our friendship will last long forever. Thanks to my next neighbour in the office, Off, for his kindness to share all kind of thoughts despite being always busy with his work.

Next, my special recognitions to my friends Linh and Kittiwet, best Ph.D buddies during my stay in Delft. To Kittiwet, despite making me sometimes feel a bit annoyed by your character, deeply inside you are a nice brother to me who always makes sure everything is under control. To Linh, I have no word to describe your kindness and help. As a good friend, I was amazed by your inner spirit that made you stronger and stronger during the hard times you faced when you were in Delft. Thank you also for preparing marvelous Vietnamese dinners and helping me to wrap my snow globes during your short visit in Delft last year.

My great appreciation is conveyed to Dr. Suryadi, a.k.a Pak Sur who is never tired to entertain me during my stays in Delft. I believe that we have a great time discussing lots of things either related or non-related to my work, albeit sometimes I have to disagree with your ideas and opinions. It is my great pleasure knowing you and I hope that we can be more productive in publishing more papers in future.

Many thanks also to Yuli, Fiona, Gladys, Sony, Clara and Tarn for always making my days in Delft cheerful. Yuli, you are the most reliable travel agent I have ever known, who always ensures our vacation trips outside the Netherlands are well organised and enjoyable. Fiona, undoubtedly you are one of the best cake-bakers who bakes beautiful and tasty cakes for my birthdays. Although we had a hard time a while ago, that ties our good friendship until today. Tarn, thank you for guiding me to put contact lenses into my eyes. It was one of the most terrifying moments in my life.

To all Malaysian in the Netherlands, particularly Anuar and his family, Junadi and his family, Zihan and Faezah, thank you very much for being part of my family in Delft. Your presence cheers up my life and your kind assistance whenever requested make my life in Delft so comfortable. May God repay all your kindness and bless you guys. Not forgotten all administrative staff in IHE, Tonneke, Anique, Jolanda, Silvia, Marielle thank you very much for your help.

Last but not least, thanks to my loving family for always being supportive and eternally encourage me to achieve my dream. Mak and Aboh, without your blessing I would have never crossed the border to gain such a wonderful experience in the Netherlands. To my elder sister and her family, thank you very much for providing a great hospitality during my visits to your place in London. Our last vacation to the northwestern cities in United Kingdom was a great experience and certainly I had a great time with the kids. I wish you and your husband all the best in your studies. To my youngest sister, thank you for accompanying me everywhere I go during my short vacations in Malaysia. To my elder brother and younger brother, many thanks for all your supports. May Allah bless you all.

Shah
Mohd Shahrizal Ab Razak
Delft, June 9, 2015

*Netherlands Research School for the
Socio-Economic and Natural Sciences of the Environment*

D I P L O M A

For specialised PhD training

The Netherlands Research School for the
Socio-Economic and Natural Sciences of the Environment
(SENSE) declares that

Mohd Shahrizal Bin Ab Razak

born on 27 May 1983 in Pekan, Malaysia

has successfully fulfilled all requirements of the
Educational Programme of SENSE.

Delft, 9 June 2015

the Chairman of the SENSE board

Prof. dr. Huub Rijnaarts

the SENSE Director of Education

Dr. Ad van Dommelen

The SENSE Research School has been accredited by the Royal Netherlands Academy of Arts and Sciences (KNAW)

K O N I N K L I J K E   N E D E R L A N D S E
A K A D E M I E   V A N   W E T E N S C H A P P E N

The SENSE Research School declares that Mr M.S. Ab Razak has successfully fulfilled all requirements of the Educational PhD Programme of SENSE with a work load of 44 EC, including the following activities:

SENSE PhD Courses

- Environmental Research in Context (2011)
- Research in Context Activity: Co-organising meeting and proceedings on Preparedness for Netherlands Centre for Coastal Research days, UNESCO-IHE, Delft (2014)

Other PhD and Advanced MSc Courses

- NCK Summer School: Estuarine and Coastal Processes in relation to Coastal Zone Management, TU Delft, The Netherlands (2011)
- Xbeach Basic - Advance course, Deltares, The Netherlands (2011)

Management and Didactic Skills Training

- Supervising MSc student with thesis entitled 'The effect of wave chronology on the shoreline response to the submerged breakwater' (2013)
- Member of the UNESCO-IHE PhD Association Board (PAB) (2011-2015)
- Supporting the organisation of scientific external evaluation of UNESCO-IHE and the SENSE Research School (2014)

Oral Presentations

- *Determination of equilibrium stages of headland bay beaches on the east coast of Malaysia.* River Coastal and Estuaries Morphodynamics, 6-8 September 2011, Beijing, China
- *Sand bypassing and shoreline evolution near coastal structure, comparing analytical solution and XBeach numerical modelling.* International Coastal Symposium, 8-12 April 2013, Plymouth, United Kingdom
- *Headland impacts on surf zone current circulation.* International Coastal Symposium, 13-19 April 2014, Durban, South Africa

SENSE Coordinator PhD Education

Dr. ing. Monique Gulickx

Printed and bound by CPI Group (UK) Ltd, Croydon, CR0 4YY

21/10/2024

01777101-0009